BEYOND BREAKING EVEN

The Service Business Owner's Playbook for Real Profit

Justin Hubbard

COPYRIGHT

Beyond Breaking Even: The Service Business Owner's Playbook for Real Profit

Published by Hubb & Spoke Media

First Edition, 2026

ISBN: 979-8-9953907-0-1

Copyright Registration Pending

Printed in the United States of America

For permissions, bulk orders, or speaking inquiries: Justin@HaulingHubb.com

HaulingHubb.com

DEDICATION

For my incredible wife and our three (soon to be four) beautiful children — this book is dedicated to you.

To my wife, whose unwavering support has been the foundation on which I've built my dreams. Your knack for planting seeds of ideas so subtly in my mind that I awaken to them as if they were my own is nothing short of magical. You are the quiet strength behind every word written.

To my three little girls — you are the reason the work I do in the early mornings and late nights is not just bearable, but filled with purpose. Each moment of fatigue is outweighed by the laughter and light you bring into my life. You are my motivation, reminding me daily of why every effort is worth it, and why the journey matters just as much as the destination.

ACKNOWLEDGEMENTS

I've learned — and continue to learn — from so many sharp minds, both well-known and unknown, who had the courage to put their thoughts, lessons, and processes into print. Books have been my teachers ever since high school, which is funny because I hated reading back then. I never cracked a book unless I had to. After high school, that changed — and I owe much of my growth to the people who took the time and the risk to share what they knew.

And it's not just books. We live in a time when education is everywhere. Newsletters, courses, podcasts, long-form blogs — real creators and real educators putting out content every day have shaped how I think and how I build. Add in the endless stream of free information online, and now with tools like LLMs, it's easier than ever to grab that knowledge and put it into practice.

This book is a culmination of all those influences — thousands of different pieces I've picked up, tested, and stitched together. The information is out there for anyone with a smartphone in their pocket. The difference comes down to whether you're willing to go after it, learn from it, and actually use it.

Less entertaining. More self-educating.

FOREWORD

By Tony Nava

Starting a business isn't easy. Nobody tells you how lonely it can feel when you're out there trying to figure it all out on your own. You make decisions you second-guess for months. You question yourself more than anyone else ever could. And when you need help, there's usually nobody to ask. Most of us just wing it and hope we don't crash too hard.

That's where a lot of first-time entrepreneurs get stuck. We turn to YouTube or some online expert hoping for real answers and all we find are vague promises that don't translate to anything useful. I fell into that same trap. I made mistakes, wasted money, and learned the hard way.

Then I met Justin. He didn't come at me trying to sell something or convince me how much better he was than everyone else. He just asked for a chance to help me grow. That right there told me everything I needed to know about him.

Working with Justin completely changed how I look at business. He helped us scale our Google Ads — but that was just the start. He became a mentor, a coach, and someone who actually gives a damn. He helped me grow not just the company but my mindset. We went from low six figures to over a million dollars in a single year.

But the money isn't the main thing. It's the confidence. Knowing that I built something real — something that works — because of what I learned from him.

Justin doesn't just talk business. He lives it. What he shares in this book isn't theory. It's tested. It's real. He cares about the people he helps, and in this industry, that's rare.

If you're reading this, understand something — this life isn't for people looking for shortcuts. It's for the ones willing to put in the work, take risks, and build something that lasts. That's what this book teaches.

Come on. Let's get it.

Tony Nava — Founder, Texas Junkers (Est. 2021) & Houston Stucco Solutions (Est. 2025) Houston, TX | U.S. Marine Corps Veteran | Seven-Figure Operator

PRAISE FOR BEYOND BREAKING EVEN

"I've spent time and had enough real conversations with Justin to know this isn't theory, it's how he actually operates. He's built a business that runs without him, which is something most owners talk about but never pull off, either because they don't have the guts or the know-how. That doesn't happen by accident, it's built on systems, processes, and discipline. What he teaches is the antidote to vague business advice and shows what it really takes to build something that doesn't rely on you to survive."

Casey Walsh – Founder & CEO, Stand Up Guys Junk Removal, GA | Est. 2008 | $10M+ Annual Revenue

---

"Justin writes like a fellow entrepreneur who's lived through it. His straightforward approach, backed by real experience, makes complicated financial and operational strategies simple and actionable. You can tell he's been in the trenches — and that's what makes this book stand out."

Barry Hartman — Co-Founder & CEO, 505-Junk, BC | Est. 2011 | $6M+ Annual Revenue

---

"Justin Hubbard is someone who doesn't just talk about the future — he builds with it. His attention to emerging trends like AI, paired with real-world success in the dumpster industry, makes his insights both forward-thinking and grounded in practical experience."

Jake Still — Co-Founder, Junk Rescue, NJ | Est. 2016 | $4M+ Annual Revenue

---

"Beyond Breaking Even is my story in action. When I met Justin, I had a pickup, a trailer, and four dumpsters. Today I've got a hook-lift truck, thirty dumpsters, and a second truck on the way. His strategies helped me more than double my business and gave me the clarity to run it with confidence."

Leo Falardeau — Owner, Drop N Go Haulers, MA | Est. 2022 | $500K+ Annual Revenue

---

"In an industry flooded with get-rich-quick schemes and revenue braggarts, Justin cuts through the noise with honesty and serious expertise. This isn't another flashy course promising millions — it's a real roadmap from someone who's actually built a profitable business. This book doesn't just show you how to run a service business. It shows you how to think like an entrepreneur in any industry."

Matt Beasley — Co-Owner, Aloha Junk Man, HI | Est. 2018 | $1.6M+ Annual Revenue

---

"I consider Justin to be the foremost expert in the dumpster rental industry. He's been in the game longer than almost anybody, adapted to change, implemented new technology, and — most importantly — structured his business so it can operate entirely without him. Even if that's not your goal, it's important to build a business that can run without you. Justin has mastered this, and I can speak from experience when I say it's much easier said than done."

Bailey Stewart — Owner, Blue Bin Dumpster, CO | Est. 2024 | $550K+ Annual Revenue

---

"Justin has more wisdom in the service industry than most people

who've been in the game longer than him. I've been in the hauling and waste industry for over ten years, most of it at a high level — and I've learned a lot from his unique, common-sense approach to business. This book is a tangible roadmap for fine-tuning existing operations or laying the right foundation for new ventures."

Brett Robinson — Owner, Greenwave Waste Solutions, AZ | Est. 2025 | 10+ Years Industry Experience

---

"Justin writes like he's right there growing the business with you. His experience and no-nonsense approach make this book both relatable and actionable."

Rob Paradis — Founder & President, Junk Bear, CT | Est. 2019 | Multi-Seven-Figure Operation

---

"Justin breaks down complex marketing strategies into clear, practical steps you can implement immediately. His real-world insights and straightforward style make this a must-read for any service business owner looking to drive results."

Kyle McAnaugh — Owner, Junk Rescue, AZ | Est. 2017 | Seven-Figure Operation

---

"Reading this feels less like generic advice and more like personalized guidance from someone who's actually been there."

Travis Lippolt — Co-Owner, VanGo Junk Removal, NY | Est. 2015 | Seven-Figure Operation

---

"Justin writes from true, successful experience — zero theoretical anecdotes. He shows service owners how to move from survival mode to building a business that gives back time, profit, and free-

dom."

Eamonn Duignan — Founder, Green Coast Rubbish, BC | Est. 2006 | Seven-Figure Operation

---

"A must-read for any service business owner. Justin gives you pragmatic guidance you can use immediately. The best business book I've read in a long time."

Coleman Clark — Owner, Clarks Junk Removal, RI | Est. 2023 | $400K+ Annual Revenue

---

"This book is rare — it blends lived experience with actionable strategies. Hubbard not only explains what to do, he shows why it matters and how to make it work in the real world."

Mike Linebarger — Co-Owner, Exhale Junk Removal, PA | Est. 2023 | $250K+ Annual Revenue

---

"Just when I thought we were going to give up, I stumbled across one of Justin's newsletters. Before I knew it we were emailing back and forth. I have learned so much from this incredible person. Beyond Breaking Even is a must-have — it will not only help you grow and scale your company, but take it to places you didn't think were possible."

Erick R. Mendoza — Owner, Texas Junk Brothers, TX | Est. 2025 | Startup

---

"Justin is an entrepreneur you want to learn from. He started from the bottom and made it. He's genuinely passionate about helping others."

Jeffrey Beasley — Owner, Strongsons Junk Removal, CA | Est. 2019 | $1.1M+ Annual Revenue

---

"He doesn't just teach business — he's built businesses, struggled through the same challenges, and figured out what actually works. This book feels like having him in your corner, showing you the direct path to profit."

Abe Chehade — Owner, Freedom Junk Hauling, CA | Est. 2024 | Corporate to $250K+ Annual Revenue

---

"Getting advice from Justin — someone with real-world experience and the scars to prove it — helped me run my company more efficiently and profitably, without any of the usual crap. If you own a service-based business, this book will pay for itself a hundred times over."

Royce Porkert — Owner, Junk Holler, ID | Est. 2012 | $500K+ Annual Revenue

---

"Justin's book is modern, relevant, and truly needed. It's perfect for anyone seeking clear direction on how to run and grow a business. His personal experience is proof that the steps he outlines genuinely work."

Taneesha Maxwell — Owner, T Maxwell Junk Removal & Cleanouts, PA | Est. 2018

"This book hits every stage of the business journey — from the basics you need when just getting started to the challenges we all face when it's time to scale. If you want to grow, make real money, and build a business that runs like a machine, you need systems,

guidance, structure, and mentorship. This book delivers all of that. Justin's the real deal."

Mario Ortega — Owner, 2Mexicans Junk Removal, TX | Est. 2023 | $400K+ Annual Revenue

---

"Justin crafted a real-world guide to building a business in the service industry. You can tell it's written by someone who's actually been there before. Every page is packed with practical insight, hard-earned lessons, and strategies that actually work. It's an invaluable resource for anyone serious about starting or scaling in this space."

Mike Ferrari — Owner, All Out Junk Removal & Demolition, FL | Est. 2022 | Multi-Seven-Figure Operation

---

"Justin is one of a kind in an industry full of people who take more than they give. He sees the bigger picture when it comes to collaboration and supports young entrepreneurs behind the scenes with very little credit. Justin, thank you for being part of our journey and showing us what it looks like to give back to others. We hope to lead by your example for as long as we're in business."

Tanner Hurst — Co-Founder, JT Junk Solutions | Est. 2022 | $300K+ Annual Revenue | 349,000 Social Followers

---

CONTENTS

Preface | *Page 17*

How To Get The Most Out Of This Book | *Page 20*

Introduction | *Page 22*

Part 1: Foundations & Money

Chapter 1: Your P&L Is Lying to You | *Page 25*

Chapter 2: You Can Be Profitable and Still Go Broke | *Page 43*

Chapter 3: What Does a Job Actually Cost You? | *Page 51*

Chapter 4: Plug the Leaks | *Page 58*

Chapter 5: Stop Guessing What to Charge | *Page 73*

Chapter 6: The Tax Trap Most Walk Right Into | *Page 83*

Part 2: Market Positioning

Chapter 7: Secret Shopping | *Page 94*

Chapter 8: Stop Chasing Every Job | *Page 102*

Chapter 9: Become the Name They Think of First | *Page 111*

Chapter 10: One Bad Review Can Bury You | *Page 121*

Chapter 11: Your Best Lead Is Already a Customer | *Page 133*

Chapter 12: Where 80% of Your Money Is Coming From | *Page 143*

Part 3: Marketing & Lead Flow

Chapter 13: Your Website Is Either Selling or Sleeping | *Page 151*

Chapter 14: Own Your Zip Code on Google | *Page 164*

Chapter 15: Every Job Site Is a Billboard | *Page 176*

Chapter 16: Stop Winging It When They Call | *Page 186*

Chapter 17: The Money Is in the Follow-Up | *Page 201*

Chapter 18: Hooking Clients on Repeat | *Page 212*

Part 4: Operations & Efficiency

Chapter 19: Run Tighter Routes, Make More Money | *Page 221*

Chapter 20: SOPs — Getting Out of Every Job | *Page 231*

Chapter 21: The Tools That Buy Back Your Time | *Page 240*

Chapter 22: Your First Hire Will Make or Break You | *Page 250*

Chapter 23: Build a Team That Doesn't Need Babysitting | *Page 262*

Chapter 24: Pay Less for Everything | *Page 273*

Part 5: Expansion & Innovation

Chapter 25: Strategic Partnerships That Print Money | *Page 282*

Chapter 26: Hidden Revenue Streams | *Page 290*

Chapter 27: The Recurring Revenue Playbook | *Page 299*

Chapter 28: One Accident Can End Everything | *Page 307*

Chapter 29: Build It Like You're Going to Sell It | *Page 318*

Part 6: Future-Proofing & AI

Chapter 30: AI Isn't Coming — It's Already Here | *Page 331*

Chapter 31: How to Use AI Before Your Competitor Does | *Page 343*

Chapter 32: The Employee That Never Calls Out Sick | *Page 353*

Chapter 33: Your First Digital Employee | *Page 361*

Chapter 34: The Future of Being Found | *Page 370*

Chapter 35: Your Next Competitor Used to Wear a Tie | *Page 383*

Chapter 36: History Rhymes | *Page 393*

Closing

Chapter 37: What Winning Actually Looks Like | *Page 406*

Epilogue | *Page 415*

About the Author | *Page 416*

Glossary | *Page 418*

Appendix A: The Business Metrics Scorecard | *Page 432*

Appendix B: Job Cost Worksheet | *Page 445*

Appendix C: Break-Even Calculator | *Page 450*

Appendix D: Recommended Tools and Resources | *Page 455*

PREFACE

Most people are trapped and don't even know it. They wake up, punch the clock, trade time for money, and call it a life. I was never built for that. Business — real entrepreneurship, creating something out of nothing — has always been my way out. It's how you buy back your time, your freedom, and live life on your own terms.

Freedom comes in a few forms, and business is the key to unlocking them:

Financial freedom: earning enough that money stops being a source of stress and starts being a tool.

Time freedom: controlling your own schedule. No begging for days off — you decide when to work and when to rest.

Locational freedom: running your business from anywhere, or not at all.

Relational freedom: choosing who you spend your time with. No more bosses you hate or toxic people you can't escape.

Freedom from conditioning: breaking away from the system that's designed to keep you on the wheel.

And what most people miss is that business, when done properly, is your way out. Your income isn't tied to inflationary raises or how many years you've warmed a seat somewhere.

Business is the unlimited income hack.

You — someone from a small town — can build something that puts you in the top 1% of earners in this country. No job could ever do that, especially not in just a handful of years. That's the brilliance of business done well: business with profit. Relentless profit focus.

And look at me — I wasn't some star student on the fast track. In high school I graduated with a C average, just enough to stay eligible for baseball. A couple colleges recruited me, but only one gave me a shot, and only if I did a semester at community college first. I got in. And once I was there, it was the same as high school — I went to class some of the time, but I couldn't focus. I'm not a standardized learner.

And I paid for it. Dearly. I walked out of school with over $120,000 in debt at a double-digit interest rate, into a brutal job market with no prospects. That bill was mine to carry — but it was also part of the conditioning. We're told to go to college, rack up debt, and then go work for somebody else. Most schools are training employees, not nurturing self-starters ready to take on the world.

And the real kicker? Banks and institutions will line up to give you six figures in loans for a degree that might not pay off. Practically rubber-stamped, no questions asked. But ask for a fraction of that to start your own business? Suddenly you're deluded, risky, unproven — and shown the door. That tells you everything about the system and who it's designed to serve.

And by the way — I paid those loans off 19 years and 3 months early! Not from a job. From building my own income. From one truck to multiple businesses — I've built from scratch, acquired two hauling companies, and started and sold a gutter cleaning business in three years.

It took years of building businesses to unlearn the indoctrination. To realize the real education is out there if you're hungry enough to find it. Everything I know in business came from reading, testing, failing, and trying again.

This book came out of countless conversations I've had with haulers and other service business owners who are working their asses off year after year and still struggling. A lot of these guys are one truck repair away from being completely done for. I've seen the stress in their faces and heard it in their voices. The work ethic isn't the problem. The way the game is set up is the problem. The numbers, the margins, the constant grind with no profit at the end — that's what's keeping them trapped.

And that's why I wrote this book.

— Justin Hubbard

HOW TO GET THE MOST OUT OF THIS BOOK

This isn't a book you read once and shelve. It's a working tool — something you dog-ear, mark up, and come back to as your business grows.

Here's how I'd approach it: don't try to do everything at once. Pick one chapter that speaks to where you're bleeding right now, apply it, and see what changes. Then come back for the next one. A phased approach beats overwhelm every time.

If you've got a team, bring them in. Share the chapters that apply to what you're building together. The best ideas I've ever had came out of conversations with people in the trenches — your team is closer to your problems than you think.

And when something doesn't fit your business exactly as I wrote it — adapt it. The principles here are universal. The execution is personal. Your market, your customers, your crew — that context is yours. Apply the thinking, not just the tactic.

Come back to this book. Seriously. What hits different when you're at $200k in revenue is going to read completely differently when you're at $800k. The book doesn't change — your lens does.

And if you think I missed something, email me: Justin@HaulingHubb.com. I'm always in the trenches, always testing, always learning.

Let's get into it.

A note on tools: Service Hubb AI is a CRM platform I built specifically for service businesses, and you'll see it referenced throughout this book because it's what we actually run at Grizzly Junk Pros.

Every capability I describe it handling exists in other platforms too — I note alternatives where relevant. My goal is always to teach the principle, not sell the product.

INTRODUCTION

"The means of learning are abundant—it's the desire to learn that's scarce." —Naval Ravikant

Starting a business is the best and hardest thing you'll ever do. Nobody tells you that upfront. They show you the highlight reel—the freedom, the flexibility, the money—but skip the part where you're staring at your bank account at 11pm wondering where it all went.

It's not just about the cash you bring in — it's about what you actually get to keep. That's where most people get it twisted. You've probably heard the line, *"Revenue is vanity, profit is sanity."* I'd add one more: *"and data brings you clarity."* Chasing revenue for bragging rights while ignoring what's left after expenses is a fast track to burnout — and probably bankruptcy. Profit is the difference between grinding endlessly and building something that actually supports your life.

I've seen it too many times. Service business owners with phones ringing, trucks running, jobs booked — looks like progress on the surface. But when the dust settles at the end of the month, the numbers don't add up. All that sweat, all those hours, and the business barely breaks even. *"Where did all my money go?"* That's the question this book exists to answer.

You're going to hit dead ends. Lose money on jobs. Hire the wrong people. Spend months on the wrong strategy. That's not failure—that's the game. The difference between the ones who make it and the ones who don't is adaptability. Knowing when to cut losses before they bleed you dry. Too many owners cling to a failing path because of what they've already put in. Sunk cost thinking will kill a business faster than a slow market.

What's kept me going isn't luck — though I'll say this: the

harder you work, the luckier you tend to get. What's really made the difference is a repeatable system. Evaluate, adjust, repeat. Try things. Test them in the real world. Keep what works, cut what doesn't. That cycle is what keeps your business alive when others fold.

This is my playbook for profitability. No MBA required — just real strategies from businesses I've actually run. Some lessons came from big wins. Most came from painful losses. All of them are here so you don't have to learn the hard way.

By the time you finish, you'll know how to read your numbers and control your cash, position yourself in a crowded market, build real lead flow instead of relying on hope, tighten your operations until efficiency becomes a competitive edge, expand without overextending, and future-proof your business for the AI-driven world we're already in, and apply the five rules that govern what you do with profit once you have it — because building it and keeping it are two different skills.

I also write every week in the Haulers' Edge Newsletter — breaking down what I'm testing, what's working, and what's not. Real time, no filter. If you want to keep leveling up after this book, that's where to find me.

Now let's get your business out of survival mode.

—Justin Hubbard

PART 1: FOUNDATIONS & MONEY

CH. 1 YOUR P&L IS LYING TO YOU

> *"Accounting is the language of business." — Warren Buffett*

Most service business owners are flying blind. They know revenue is up, they know they're busy, and they assume that means things are good. Then the end of the month hits and the bank account tells a completely different story.

If you already review your P&L monthly, good — you're ahead of most. But here's what experienced operators still get wrong: reading the bottom line as a health signal instead of using the document as a diagnostic tool. This chapter isn't about learning what a P&L is. It's about reading it in a way that tells you what to change.

That gap — between what you think is happening and what the numbers actually say — is where businesses quietly bleed out. Your Profit & Loss statement is the document that closes that gap. Not because it's complicated. Because most owners never learn to read it honestly.

This chapter breaks down your P&L into the parts that actually matter and shows you how to use it as a weapon, not just a report you hand to your accountant once a year.

Revenue

Revenue is the top line — every dollar that comes in before a single expense gets paid. It's the total of everything your business earns across every service you offer.

For a hauling or junk removal operation that might include residential junk removal, commercial cleanouts, dumpster rentals, demolition services, scrap and recyclable sales, labor-only jobs, donation partnerships, and specialty disposal. For a landscaper it's mowing contracts, installs, and seasonal clean-ups. For a cleaning company it's recurring residential, one-time deep cleans, and commercial accounts. The categories look different — the principle is the same. Every service line feeds the top of your P&L, and they don't all earn equally.

Here's a simple example. Say you brought in $10,000 last month: $6,000 from junk removal, $3,000 from dumpster rentals, $1,000 from salvaged metal sales. That's your revenue — every dollar in the door before anything goes out.

But here's what most owners miss. Revenue doesn't tell you what you made. It tells you what came in. A busy month with weak margins can leave you with less actual profit than a slower month where you ran tight. That's the trap. And that's what the rest of your P&L is for.

Hidden Revenue You're Leaving on the Table

Every service business has money it isn't capturing — add-ons, secondary services, materials fees that represent real effort and real cost but never show up on an invoice because nobody built them into the pricing.

The question isn't whether these opportunities exist. They do. The question is whether chasing them is worth what they cost to deliver. When you're running solo, extra revenue streams often make sense. When you're paying a team to execute, every add-on needs to earn more than it costs in time, labor, and management attention. Some won't. Audit them honestly before you scale them.

The hidden lines most service businesses miss:

Environmental or recycling fees charged per appliance or tire hauled — standard in many markets, often forgotten at quoting time

Fuel surcharges applied to jobs beyond a set radius or above a fuel price threshold

Minimum trip charges that protect you from sub-threshold jobs that eat time for almost no margin

After-hours or same-day premiums for customers who need it fast

Item-level add-on pricing for heavy or specialty items like safes, pianos, or concrete that carry real extra cost

None of these require new services. They require building the fees you're already absorbing into the number you give the customer.

Direct Costs: What It Actually Costs to Do the Job

Direct costs — also called cost of goods sold or cost of services — are the expenses tied directly to completing a job. When work picks up, these go up. When work slows, they come down. They only exist because the job exists.

For most service businesses that means labor, fuel, disposal or dump fees, materials and supplies, and any subcontractor costs. The exact line items differ by trade. The discipline of tracking them doesn't.

Here's a concrete example. You charge $600 to clean out a garage. To get it done: $50 in dump fees, $30 in fuel, $120 to a day laborer, $20 in supplies and gear. That's $220 in direct costs. Subtract that from $600 and you've got $380 — your gross profit on that job. Your gross margin is 63%. Not bad. It means 63 cents of every dollar from that job is left after the immediate costs of doing it.

Now run that same math across every job last month. You'll start seeing which services are actually making you money and which ones just feel like they are.

The most common mistake here is underestimating direct costs when quoting. A job that looks profitable on paper can shrink to almost nothing once every variable is accounted for. Build a checklist or use a pricing tool so nothing gets missed. In my company we use the Haulers' Edge AI tool — upload job photos, add notes, and it returns a price range, margin estimate, and time forecast. Whether you're using AI or a spreadsheet, the point is the same: never quote from gut feel.

Two other things worth watching. First, services that feel like they're adding value — like special recycling handling or material sorting — still come with time and expense attached. If the return doesn't cover those costs, either build them into the price or limit how often you offer them. Second, variable costs like fuel and disposal fees fluctuate. Review them regularly and adjust pricing when they move. Don't let costs creep up while your prices stay flat.

Overhead: The Costs That Never Stop

Overhead — also called indirect costs or operating expenses — is what it costs to keep your business running whether you're on a job or not. Insurance, vehicle payments, software subscriptions, marketing, admin wages, utilities. These don't rise and fall with individual jobs. They're the floor your business sits on every single month.

That's what makes overhead dangerous. No single line looks like much. But according to a U.S. Bank study, 82% of small business failures are tied to cash flow and cost management problems — not lack of revenue *(U.S. Bank, 2019)*. Overhead is usually where it starts.

That $30 software subscription you forgot about is $360 a year. Stack five of those and you've quietly lost $1,800. That marketing campaign you kept running out of habit — when did you last check how many leads it actually produced?

Track what every dollar of overhead is doing for you. Cut what isn't producing. Double down on what is. And when the business starts growing, resist the urge to upgrade everything at once. New trucks, bigger equipment, expanded staff — these raise your floor. If revenue doesn't grow faster than overhead, margins shrink. I learned this firsthand. The data at the time supported adding new equipment — the numbers looked right. But a shift in the economy and in consumer behavior changed the timeline entirely, and it took us 18 months longer than planned to get that asset into full production. For 18 months we carried the cost with restricted cash flow. Not a catastrophic mistake, but an expensive one that limited our ability to operate the way we wanted. The data can be right and the timing can still be wrong.

The real discipline isn't about waiting for a specific moment — it's about the difference between evidence and optimism. Buying because your current operation is genuinely constrained and the numbers confirm it is a different decision than buying because you're excited about where the business is going and you're projecting it will catch up. One is disciplined growth. The other is how you end up carrying payments on equipment that's sitting idle. If lead times mean you need to order months before peak demand, that's fine — factor it in. What you're avoiding is buying capacity your operation hasn't earned yet. The goal is lean, not bare. Keep your setup one where every new dollar of revenue adds more profit, not just more bills.

Some overhead is non-negotiable — insurance, reliable tools, essential software. That's not the problem. The problem is optional overhead that accumulates without scrutiny. Review it quarterly at minimum. Catch it before it snowballs.

Your Break-Even Point: The Number That Actually Matters

Before you can go beyond breaking even, you need to know exactly what breaking even looks like for your business. This is a number most service business owners have never calculated — and it shows.

Your break-even point is the amount of revenue you need to bring in every month before a single dollar of profit exists. The standard formula is straightforward:

Break-Even = Total Monthly Overhead ÷ Gross Profit Margin

Here's how it works in practice. Say your monthly overhead — insurance, truck payment, software, marketing, phone, everything fixed — adds up to $8,000. And your average gross profit margin across all jobs is 55%. Divide $8,000 by 0.55 and your break-even is roughly $14,500 in monthly revenue. Every dollar above that is profit. Every dollar below it is a loss.

But there's a layer most operators miss. Some costs in a service business only exist because jobs exist. Subcontractor labor you only pay when work comes in. Disposal fees that scale with volume. Fuel that increases when the trucks are running hard. These aren't truly fixed overhead and they aren't direct job costs either — they're variable costs that rise and fall with your revenue.

If your business has significant costs in this category, the standard formula understates your real break-even. Those variable costs are quietly consuming margin that the formula assumes is flowing toward profit. The more accurate version accounts for them:

Break-Even = Fixed Overhead ÷ (Gross Profit Margin – Vari-

able Cost Rate)

If your fixed overhead is $8,000, your gross margin is 55%, and your variable costs run about 10% of revenue, your break-even becomes $8,000 ÷ 0.45 — roughly $17,800. That's $3,300 higher than the basic formula suggests. For an operator running near break-even, that gap is the difference between thinking you're profitable and actually being profitable.

The specific formula matters less than the discipline of knowing your number honestly. Add up your true fixed costs. Calculate your real gross margin from your actual job data. Account for the costs that scale with your volume. Then divide.

Run this number for your business right now. If you don't know it, you don't know whether a busy month actually made you money or just kept the lights on. A lot of operators are running hard every day and never clearing that line. That's the whole problem this book exists to solve.

And as your business grows — you add a truck, hire a driver, take on a new lease — recalculate immediately. Every new fixed cost raises your floor. Every new variable cost tightens your margin. You need to know by how much.

One more dimension worth understanding: your break-even number is fixed but your ability to clear it isn't. Most service businesses have seasons — dumpster rentals spike in summer when contractors are active and homeowners are tackling projects, and slow in winter when neither group is moving. Landscapers, cleaners, pressure washers, and most home service trades follow similar patterns.

Your break-even doesn't move with the seasons. Your overhead is the same in January as it is in July. What changes is how many months you're comfortably above the line versus how many months you're below it or barely clearing it.

This is why calculating break-even on an annual average can

be misleading. If your business generates $25,000 in revenue during peak months and $8,000 during slow ones, an average month looks fine — but you may be running at a loss for several months a year and surviving on the surplus from the busy season rather than on a consistently profitable operation.

The more useful practice is to run your break-even calculation against your slowest realistic month, not your average month. If your fixed overhead is $8,000 and your slowest month reliably brings in $10,000, you have $2,000 of margin above break-even even in your worst month. That's a stable business. If your slowest month brings in $6,000, you're losing $2,000 that month regardless of how well summer went — and that loss has to be funded from somewhere.

Two things protect you here. A cash reserve built during peak months specifically to carry the slow ones without stress — we'll cover cash reserves in Chapter 2. And the recurring commercial revenue from Chapter 27 — contractors, property managers, and commercial accounts that generate work in every season rather than concentrating it in the warm months. Both of those strategies exist partly to smooth the seasonal swings that make break-even a moving target in practice even when the formula stays the same.

Know your slow month number. Build toward covering it without relying on summer to bail you out. That's what a stable service business looks like across a full calendar year.

Debt: Borrowed Money Is Tomorrow's Profit Spent Today

Taking on debt isn't automatically a bad move. Smart debt — borrowing that generates more revenue than it costs — can accelerate growth. The problem is dumb debt, and it's more common than anyone admits.

The trap most service business owners fall into is borrowing ahead of demand. Buying trucks, containers, or equipment before the work exists to support them means making payments whether or not the asset is earning. Your P&L might still look profitable while your bank account quietly empties. Profit is an accounting measure. Cash flow is what actually pays your bills. *(More on that distinction in Chapter 2.)*

Watch your interest expense line closely. That number is the real cost of every loan on your books — and it doesn't adjust when business gets harder. If your costs are primarily variable — labor, fuel, disposal fees that scale with jobs — more revenue generally means more profit and fixed debt becomes easier to carry over time. That's the healthy version of growth. The danger is when you take on new fixed costs to chase revenue that doesn't fully materialize. A second truck, a new hire on salary, an expanded facility — these obligations exist whether the jobs show up or not. If you borrowed to fund that expansion and the revenue growth lags behind, the loan payment takes the same dollar amount out of your account every month regardless of what the business earned. Debt tied to variable cost operations tends to get easier with growth. Debt tied to fixed cost expansion can strangle you if the growth doesn't come. Know which one you're carrying.

The rule: let growth drive purchases, not the other way around. Only borrow when it directly creates revenue or protects the business from a larger loss. And never use debt to cover operating expenses — payroll, utilities, day-to-day costs. Those expenses disappear the moment you pay them. The debt doesn't. That road ends one way.

If debt becomes necessary to bridge a slow season, treat it as temporary and have a specific repayment plan before you sign. The smarter move is building cash reserves during busy stretches so you're never forced into survival borrowing in the first place.

Refinance when better rates are available. Pay down high-interest balances first — credit cards and merchant cash advances especially, which can carry effective rates north of 40% when you run the real math *(Federal Reserve, 2023)*. And always ask one question before borrowing: will this debt pay for itself? If the honest answer is no, it's probably not worth it.

Debt isn't the enemy. Mismanaged debt is.

Taxes: Keep More of What You Earn

(Chapter 6 covers taxes in full. This section gives you the essentials — don't skip it.)

The goal with taxes isn't avoidance. It's making sure you never hand over more than you legally owe — and never getting caught off guard when the bill arrives.

Start with the basics. Track every legitimate business expense: fuel, maintenance, tools, marketing, a portion of your home office. The more accurately you record, the lower your taxable profit, which directly reduces what you owe.

Set aside 25–30% of your net profit throughout the year and treat it as untouchable. Tax season should be a payment, not a crisis.

For bigger purchases, look at Section 179. It allows you to deduct the full cost of qualifying equipment in the year you buy it rather than depreciating it over time — the timing of an equipment purchase can shift your tax bill significantly. Chapter 6 breaks this down with numbers. (State rules vary — verify with your tax advisor before counting on it.)

Structure matters. A sole proprietorship leaves you personally exposed. An LLC or S-Corp election can offer both liability protection and meaningful tax efficiency. The right setup depends on your revenue level and how you pay yourself — Chapter 6 breaks this down in detail.

The tax code was written with business and property owners in mind, not W2 employees. Use it.

Where Your Own Paycheck Fits In

This is one of the most misunderstood parts of the P&L for small business owners, so let's clear it up now.

How you pay yourself depends on how your business is structured. If you're a sole proprietor or single-member LLC, your pay comes out as an owner's draw — money you pull from the business bank account. That draw doesn't show up as an expense on your P&L, which means your profit looks higher than what you're actually keeping. You need to account for what you're paying yourself when you're evaluating whether the business is truly profitable.

If you've elected S-Corp status, you pay yourself a reasonable salary, which does show up as a payroll expense on your P&L. That salary reduces your taxable business income, which is one of the main reasons the S-Corp election makes sense at higher revenue levels.

The practical takeaway: if you're an owner-operator who isn't accounting for your own labor in the numbers or what it would cost to hire someone to do your job, your margins are lying to you. A job that looks profitable might not be if your time were priced honestly. Build your own pay into how you evaluate the business from day one.

How Your Books Are Kept: Cash vs. Accrual

When you set up accounting for your business you'll encounter two methods — cash basis and accrual. Most small service businesses run cash basis, and that's probably right for you. Here's the difference.

Cash basis means you record money when it actually hits your account and expenses when you actually pay them. Simple, straightforward, and easy to manage.

Accrual means you record revenue when it's earned and expenses when they're incurred, regardless of when money actually moves. A landscaper who invoices a client in December but gets paid in January records that revenue in December under accrual, January under cash basis.

For most service businesses doing under a few million in revenue, cash basis keeps things clean and gives you an accurate picture of what's actually in your pocket. When you eventually work with a CPA — and you should — they'll likely ask which method you're using. Know the answer.

Timing Income Around Year End

Cash basis also gives you some legitimate flexibility in how you time income and expenses around year end. If you complete a job in late December and the customer hands you a check, you're not required to deposit it before December 31st. If that check sits in your desk until January 2nd, that income lands in the following tax year under cash basis accounting. Meanwhile every expense you paid in December — fuel, disposal fees, supplies, software subscriptions — counts against your current year income. The result is a lower adjusted gross income for the current year without anything aggressive or questionable. It's simply timing.

Here's where that timing can make a material difference. If you know your numbers and you can see in November that your taxable income is hovering near the top of a lower tax bracket, a few well-timed moves can keep you from crossing into the next bracket and owing a higher rate on every dollar above that line. Deferring a December payment to January, accelerating a planned January expense into December, or pulling forward an equipment purchase you were going to make anyway — any one of these can shift enough income or expense to keep you in the lower bracket for the current year.

The tax savings on that decision can easily run into thousands of dollars. Not from aggressive maneuvering — just from knowing where you stand and acting before the calendar flips.

Credit Cards as a Deduction Tool

Credit cards add another useful dimension to this under cash basis accounting. When you charge a business expense to a credit card, the deduction applies in the year the charge hits — not the year you pay the card balance. If you put $15,000 in equipment, supplies, or business expenses on your card in December and don't pay that bill until February, every dollar of those purchases is deductible in the current tax year. The IRS treats the charge date as the payment date for cash basis taxpayers. This means a credit card is effectively a tool for pulling deductions forward into the current year without requiring the cash to leave your account until later. Combined with the income deferral and expense acceleration strategies above, a business owner who knows their numbers in November has real options for managing their tax position before the year closes.

Prepaying Expenses

The same logic works in reverse for expenses. If you know you'll need supplies, equipment, or services in early January, paying for them in late December pulls those deductions into the current tax year. Prepaying certain business expenses before year end — insurance premiums, software annual subscriptions, office supplies — is a standard and legal way to reduce your current year taxable income.

A few other legitimate year-end moves worth knowing about. Accelerating equipment purchases before December 31st allows you to take the Section 179 deduction in the current year rather than depreciating the asset over several years. If you've been thinking about a new truck or a piece of equipment, the timing of that purchase can have a meaningful tax impact. Your CPA can run the numbers on whether accelerating it makes sense for your specific situation.

None of this is aggressive tax avoidance. It's the basic tax planning that every small business owner should understand and that most never learn until they're already paying more than they need to. The key is working with a CPA who understands service businesses and proactively brings these conversations to you before December 31st — not after, when the options are already gone.

Gross vs. Net: The Two Numbers That Actually Matter

Revenue tells you what came in. Margins tell you how efficiently you're turning it into money you actually keep.

Gross profit margin = (Revenue – Direct Costs) ÷ Revenue. This is what's left after paying for the job itself, before overhead takes its cut.

Net profit margin = Net Profit ÷ Revenue. This is what's left after everything — direct costs, overhead, interest, and taxes.

Most small service businesses land between 10–20% net margin *(IBISWorld, 2023)*. That means for every $100 in revenue, you're keeping $10–$20 after all bills are paid. If that number is shrinking year over year, your expenses are outpacing revenue — and it will eventually catch up with you.

Margins let you compare your services on equal footing. A

service that keeps you busy but runs thin margins is working harder for less. A quieter service with strong margins might be your actual profit engine. The P&L shows you which is which — if you're reading it honestly.

Using Your P&L as a Strategic Weapon

Your P&L isn't paperwork. It's the GPS for your business — showing you where you've been, where you are, and where to steer next.

Review it monthly. Not quarterly, not when your accountant sends it over. Monthly. Consistency is what turns raw data into real insight. You'll start catching patterns — revenue spikes in spring, fuel costs jump in summer, a service line that looks busy but never moves the profit needle.

Compare it against something. Last month. Last year. Your own targets. Numbers in isolation tell you almost nothing. Numbers in context tell you everything.

If you're handling the books yourself, consider having a CPA review them monthly. A few hundred dollars a month for clean, accurate numbers pays for itself the first time it catches a problem early enough to fix it. For tracking software, QuickBooks is the industry standard for small businesses. Wave is a solid free option if you're just getting started. Either one will generate a P&L automatically once your transactions are connected — which means no excuse not to be looking at this number every single month.

Then act on what it tells you. Here's what "reviewing it" actually looks like in practice — three things every monthly P&L review should check. First, compare this month's gross margin percentage to last month's and to the same month last year. If it dropped, find out why before you close the browser. Second, scan every expense line for anything that increased more than 10% month over month. Fuel, disposal, labor burden, software subscriptions — one of those moving without explanation is almost always a margin leak. Third, check whether your two or three largest service lines each cleared your minimum gross margin floor. If one didn't, it gets repriced or cut before next month. That's the review. It takes fifteen minutes. Do it on the same day every month.

The P&L shows you where the profit is hiding. Your job is to follow it.

The Third Document Most Operators Never Look At

Your P&L tells you if you're profitable. Your cash flow statement — covered in Chapter 2 — tells you if you can pay your bills. There's a third financial document your business produces that few owners ever look at and even fewer understand: the balance sheet. It tells you what your business is actually worth and whether it could survive a real hit.

A balance sheet has three parts. Assets are everything your business owns that has value — cash in your accounts, accounts receivable you're owed, equipment, vehicles, inventory. Liabilities are everything your business owes — loans, leases, unpaid invoices, credit card balances. Equity is what's left when you subtract liabilities from assets. That number is the net worth of your business. A positive and growing equity number means the business is building real value over time. A negative or shrinking one means debt is outpacing assets — which can happen even when the P&L looks fine, usually because equipment purchases are being financed faster than they're being paid down.

For most service business owners, the balance sheet matters in three specific situations. First, when you apply for a business loan or line of credit — lenders look at all three financial statements, not just the P&L, and a weak balance sheet will either kill the application or raise your interest rate. Second, when you're considering a major equipment purchase — the balance sheet tells you whether the business can absorb the debt or whether you're already leveraged to the edge of what's manageable. Third, when you're thinking about an eventual exit — any serious buyer will want three to five years of balance sheets alongside your P&L history. A business with clean, consistent balance sheets sells faster and at better multiples than one where the buyer has to guess at the underlying financial health.

You don't need to build a balance sheet manually. QuickBooks and Wave generate it automatically from the same transactions that produce your P&L — you just have to look at it. Once a quarter, pull all three documents — P&L, cash flow statement, and balance sheet — and ask three questions about the balance sheet specifically: Are my assets growing? Is my debt level manageable relative to those assets? Is equity trending in the right direction year over year? Those three questions take five minutes and tell you something your P&L never will.

Revenue is vanity. Profit is sanity. And data is clarity.

The operating margin and expense-to-revenue metrics that tell you whether your P&L is actually healthy are part of the monthly review in Appendix A.

Go Deeper

Every week I publish the Haulers' Edge Newsletter — real strategies, real numbers, and what's actually working in service businesses right now. No theory, no crap, just operator-to-operator insight delivered to your inbox every week. If this book is the foundation, the newsletter is what keeps it current as the game keeps changing. Scan the code below or subscribe free at HaulingHubb.com.

CH. 2 YOU CAN BE PROFITABLE AND STILL GO BROKE

"More businesses die from indigestion than starvation."
— David Packard

I want to tell you about a type of business owner I've seen more times than I can count. Books look decent. Jobs are rolling in. Revenue is up from last year. And then one Tuesday morning they can't make payroll.

Not because they failed. Not because they were lazy or stupid or made some catastrophic mistake. Because profit and cash are not the same thing — and nobody told them that until it was too late.

This is the chapter nobody puts at the front of a business book. It should be.

Profit Is a Number. Cash Is What Keeps the Lights On.

Your Profit & Loss statement — which we covered in Chapter 1 — measures profitability. It tells you whether your business made more than it spent over a given period. That number matters. But it doesn't tell you whether the money is actually sitting in your account right now, available to pay a bill that's due today.

Cash flow is the movement of money in and out of your business in real time. And the timing of that movement is everything.

Here's a simple example. You land a $15,000 commercial cleanout for a property management company. Big job, great margin. You do the work in early November, invoice them, and book the profit. But the property management company pays net-30. That means you won't see that $15,000 until December. Meanwhile your truck payment is due November 15th, payroll hits November 22nd, and your dump fees from the job are due on delivery.

On paper you're profitable. In your bank account you're scrambling.

That gap — between when you earn money and when you actually receive it — is where service businesses die. According to a study by Intuit, 61% of small business owners have struggled with cash flow, and 32% have been unable to pay vendors, themselves, or their employees as a direct result *(Intuit, 2019)*. Profitable businesses. Out of cash.

The Four Cash Flow Killers in Service Businesses

Understanding why cash gets tight is the first step to controlling it. In service businesses it almost always comes down to one of four things.

Slow-paying clients. Residential customers usually pay on the spot — that's clean. But the moment you start working with commercial accounts, property managers, real estate investors, or contractors, payment terms enter the picture. Net-15, net-30, net-60. Some of these clients will push to net-90 if you let them. Every day that invoice sits unpaid is a day your cash is working for their business instead of yours. Set clear payment terms upfront, invoice immediately after the job, and follow up without apology when invoices go past due.

Seasonal revenue swings. Most service businesses have a peak season and a slow season. For hauling and junk removal that's typically spring through fall, with winter slowdowns depending on your market. The trap is spending at peak-season levels and then watching overhead eat your cash reserves through the slow months. If you made great money in September but burned through it by December, January becomes a survival exercise. The fix is building reserves during busy stretches specifically to fund the slow ones — not spending every dollar the moment it arrives.

Growing too fast. This one catches people off guard because growth feels like success. You land three big new accounts, hire two people, lease a second truck. Revenue jumps. But payroll is weekly, the truck lease is monthly, and those new accounts pay net-30. You've committed to outflows before the inflows have arrived. According to the U.S. Bank study cited in Chapter 1, this pattern — expanding faster than cash flow supports — is one of the primary drivers of small business failure, even in otherwise healthy companies *(U.S. Bank, 2019)*. Growth is good. Growth funded by cash you haven't collected yet is a trap.

Debt service. Every loan payment you make is cash leaving the business whether or not you're busy that week. We touched on this in Chapter 1. A slow month plus a stack of fixed loan payments is a dangerous combination. This is why borrowing ahead of demand is so risky — those payments don't pause when revenue dips.

Your P&L Won't Show You Any of This

That's the brutal part. You can look at a perfectly healthy P&L and have no idea a cash crisis is two weeks away. The P&L measures what happened over a period. Cash flow is about what's happening right now and what's coming.

This is why your business needs three financial documents, not one. The P&L tells you if you're profitable. A cash flow statement tells you if you can pay your bills. And the balance sheet — covered below — tells you what the business is actually worth and whether it could survive a real hit.

A basic cash flow statement tracks three things: cash coming in, cash going out, and the net result — your running cash balance. Unlike the P&L, it's anchored in actual timing. Money doesn't get recorded when it's earned. It gets recorded when it moves.

Most small service businesses don't look at a formal cash flow statement. They look at their bank account and call it cash flow management. That works until it doesn't. The bank account tells you what's there now. It doesn't tell you that $22,000 in bills are due in the next ten days and only $11,000 is coming in before then.

Pull all three financial documents together once a quarter — your P&L, cash flow statement, and balance sheet. Together they tell you more than any one of them can alone.

The 13-Week Cash Flow Forecast

The tool that fixes this is a 13-week cash flow forecast. Thirteen weeks — one quarter — is short enough to be accurate and long enough to see problems before they hit.

Here's how it works. Every week you map out what's coming in and what's going out for the next 13 weeks. Revenue you expect to collect — not earn, collect — on one side. Every bill, payroll run, loan payment, and expense on the other. The difference tells you your projected cash balance week by week.

When you see a week three weeks out where outflows exceed inflows, you have time to act. Chase down a slow-paying invoice. Push a non-critical expense. Draw on a line of credit before you desperately need it rather than after. That lead time is everything.

For a service business operating under $2 million in revenue, this doesn't need to be complicated. A simple spreadsheet updated weekly does the job. The discipline is what matters, not the sophistication of the tool.

What a Healthy Cash Position Actually Looks Like

There's a rule of thumb that gets thrown around in small business circles: keep 3 months of operating expenses in reserve. For most early-stage service business owners that number feels impossible. But the direction is right even if the target takes time to reach.

Start smaller. One month of overhead as an untouchable reserve — money that exists specifically to bridge slow periods, unexpected repairs, or a client who pays late. The JP Morgan Chase Institute found that the median small business holds fewer than 27 days of cash buffer *(JP Morgan Chase Institute, 2016)*. That's one bad month away from crisis for most operators. One major truck repair away. One slow season away.

Build your reserve intentionally. When a big job comes in, set a percentage aside before you spend anything. Treat it like a tax — non-negotiable, automatic, invisible to your day-to-day spending. Over time that buffer becomes the difference between a cash crunch being an inconvenience and being an emergency.

Practical Moves to Protect Your Cash Flow

Beyond building reserves, there are specific habits that change the cash flow picture fast.

Collect faster. For residential work, collect at the time of service — no exceptions. For commercial accounts, shorten your terms wherever possible. Net-15 is better than net-30. A 2% discount for payment within 5 days is often worth it if it keeps cash moving. And invoice immediately. Every day between completing a job and sending an invoice is a day you pushed your own payment back.

Know your receivables. At any given moment you should know exactly how much money is owed to you and how old each invoice is. Letting receivables age quietly is how you end up cash poor while technically profitable.

Control your payables timing. Pay bills on their due date — not early. Holding cash as long as legitimately possible gives you flexibility. This isn't about being difficult. It's about managing your own float.

Build a line of credit before you need it. A business line of credit is a cash flow tool, not a bailout option. The best time to establish one is when business is good and your financials are strong. Banks lend to businesses that don't desperately need money. If you wait until you're in a crunch, it's too late and the terms will be worse. Set it up now, keep it unused, and have it available as an emergency bridge.

Watch the gap between costs and collections on big jobs. For large commercial projects — multi-day cleanouts, big demo jobs, extended dumpster rentals — you're often incurring costs weeks before you collect. Consider requiring a deposit upfront, especially for jobs over a certain dollar threshold. Fifty percent down before the job starts is standard in many trades and completely reasonable to request.

Speed to cash isn't just about deposits — it's about systematically closing the gap between work completed and money received at every point in your operation.

Review your accounts receivable weekly. Any invoice over

thirty days needs a follow-up call — not an email, a call. Any invoice over sixty days needs a conversation about whether this commercial relationship is worth maintaining on the terms it's currently operating under. Receivables that age quietly are the most expensive kind of cash flow problem because you don't see them draining you until the damage is done.

The Simple Cash Flow Habit That Changes Everything

Once a week — same day every week — look at three numbers. What came in this week. What went out. What's in the account right now. Then look two weeks ahead. What's due, and what's expected to collect.

That's it. Ten minutes a week. It sounds almost too simple to matter. But most of the service business owners I've talked to who've been blindsided by cash problems had one thing in common: they weren't looking at these numbers consistently. They were looking at their bank account when they needed to pay something and hoping the number was high enough.

Profit gets reported once a month. Cash flow happens every single day. Treat it accordingly.

Revenue is vanity. Profit is sanity. Cash flow is reality.

All three matter. But cash flow is the one that determines whether your business is open tomorrow. The cash position and average days-to-collect metrics that track this week to week are in Appendix A.

Go Deeper

For more on this and everything else in the book, the Haulers' Edge Newsletter goes deeper every week. Scan the code or sub-

scribe free at HaulingHubb.com.

CH. 3 WHAT DOES A JOB ACTUALLY COST YOU?

"The moment you think you know your numbers is the moment you stop checking them." —Justin Hubbard

Most service business owners can tell you what a job pays. Almost none of them can tell you what it actually costs.

Not the obvious stuff — the dump fee, the labor, the fuel. Those are easy. I mean the real cost. The full cost. Every dollar of time, overhead, wear, and risk that gets consumed the moment you say yes to a job.

That gap — between what you think a job costs and what it actually costs — is where margin disappears. Quietly, consistently, job after job, until you're wondering why a busy month still left you broke.

This chapter closes that gap.

Why Most Quotes Are Built on Guesswork

When a new service business owner quotes a job, the mental math usually goes something like this. The customer wants a garage cleaned out. I charged $500 for something similar last month. That felt okay. I'll say $500.

No labor calculation. No fuel estimate. No allocation of overhead. No accounting for drive time, the dump run, the hour it took to answer the call and schedule the job. Just a gut feel anchored to a vague memory of a previous job that may or may not have actually been profitable.

This is how most operators price. And it works — sort of — until it doesn't. Until fuel prices spike. Until the helper you brought costs more than expected. Until the dump fee at the transfer station went up and nobody noticed. Until the job that felt like $500 work turns out to be $650 worth of actual cost, and you've just paid a customer $150 to haul their junk.

Profitable businesses don't quote from memory. They quote from math.

The True Cost of a Job — Every Line That Counts

Job costing is the practice of identifying every cost associated with a specific job before you price it. Not after. Before. Here's every category that belongs in that calculation.

Direct labor. The hours your crew spends on the job multiplied by what you're paying them, including payroll taxes and workers' compensation. If you're paying a helper $20 an hour and the job takes three hours with two people, that's $120 minimum — before you add the employer-side burden, which typically adds another 15–20% on top *(IRS, 2024)*. So the real labor cost on that job is closer to $140. Most owners quote the $20 and forget the rest.

Your own time. This one gets skipped constantly and it's one of the most expensive line items in the business. If you're owner-operating, your time on a job has a cost whether you pay yourself explicitly or not. If your time is worth $75 an hour and you spend two hours on a job — driving, working, wrapping up — that's $150 of value the job needs to justify. If it doesn't, you're effectively subsidizing your own customers with your labor. Price your time or work yourself out of business slowly.

Fuel. Not just the drive to the job. The drive from your yard to the job, job to the dump, dump back to the yard or to the next job. Map the full route. At current fuel costs and typical truck consumption — most service trucks get 8–12 miles per gallon loaded — a 40-mile round trip in a heavy truck can run $25–$35 in fuel alone before you've touched a single item *(U.S. Energy Information Administration, 2024)*. On a $300 job that's 10% of revenue gone before you start.

Disposal fees. Dump fees vary by location, by material type, and by weight. If you're hauling construction debris, mattresses, electronics, or appliances, disposal costs more than standard household junk. Know your transfer station's rate card cold. Know what heavy loads cost versus light loads. Build the disposal estimate into every quote — not a rough guess, an actual number based on what you expect to haul.

Materials and supplies. Tarps, straps, protective gear, bags, zip ties, cleaning supplies. These feel like rounding errors until you track them across a month and realize you're spending $400 on supplies with no line item accounting for it anywhere in your quotes.

Equipment wear and depreciation. Your truck has a finite life. Every mile, every load, every dump run is consuming that life. A work truck that costs $60,000 and runs 150,000 miles over its useful life has a cost of $0.40 per mile just in depreciation — before maintenance, tires, or repairs *(standard straight-line depreciation calculation)*. On a 40-mile round trip that's $16 in depreciation baked into the job whether you account for it or not. Most operators don't. Then they're surprised when the truck needs a $4,000 repair and there's no money set aside for it.

Overhead allocation. Every job needs to carry a share of your fixed overhead — insurance, truck payments, software, marketing, phone, admin time. If your monthly overhead is $6,000 and you complete 80 jobs a month, each job needs to contribute $75 just to cover the cost of keeping the business running before any profit exists. If you're not building overhead into your pricing, your jobs aren't actually profitable — they're just covering their direct costs while overhead bleeds you dry month after month.

The Job Cost Formula

Pull these together and you get a complete picture of what a job actually costs:

True Job Cost = Direct Labor (with burden) + Owner Time + Fuel + Disposal Fees + Materials + Equipment Depreciation + Overhead Allocation

Gross Profit = Revenue – True Job Cost

Gross Margin % = Gross Profit ÷ Revenue

Here's what that looks like on a real job. A residential garage cleanout, quoted at $600.

Direct labor: two crew members, 2.5 hours each at $20/hour plus 20% burden = $120. Owner time: 1 hour driving and coordinating at $75/hour = $75. Fuel: 35-mile round trip in a loaded truck = $30. Disposal: one truck load, mid-weight = $80. Materials: tarps, bags = $10. Equipment depreciation: 35 miles at $0.40/mile = $14. Overhead allocation: $75.

True Job Cost = $404. Gross Profit = $196. Gross Margin = 32.7%.

That job doesn't clear a 35% floor — and the only reason you'd know that is because you included burden in the labor line, exactly as the formula requires. Without burden, the same job looks like 36% margin and you think you're fine. With it, you know you need to reprice. That's the whole point of running the real numbers.

Run this on your last ten jobs. Not hypothetically — actually run the numbers. You will find jobs you thought were profitable that weren't. You will find services you've been underpricing for months or years. You will also find some jobs that are more profitable than you realized, which tells you where to focus.

That exercise alone is worth more than any chapter in this book.

The Hidden Cost Most Operators Ignore: Drive Time

Here's the one that surprises people most when they actually track it. Drive time is labor. It is paid time — yours or your crew's — during which no revenue is being generated.

If you're paying two people for three hours on a job but the job is 45 minutes each way, you're actually paying for four and a half hours of labor to complete three hours of billable work. That 90 minutes of unpaid drive time is a cost. It needs to either be priced into the job or eliminated through smarter routing — which we cover in Chapter 19.

The same logic applies to time spent loading, unloading at the dump, waiting at the transfer station, and doing a final sweep of the job site. All of it is time. All of it costs money. None of it is free just because it doesn't feel like the "real" work.

Setting a Minimum Job Margin

Once you know what a job truly costs, you can set a floor. A minimum acceptable margin below which you don't take the job — or you reprice it until it clears the bar.

Most service businesses should be targeting a gross margin floor of 40–50% on individual jobs, with higher margins on simpler, faster work and the floor holding firm on more complex jobs that carry more risk and overhead *(industry standard for small service businesses)*.

That number needs to be specific to your cost structure, not borrowed from someone else's business. Calculate your overhead per job. Know your labor burden rate. Run your average fuel and disposal costs. Then set your floor based on math, not gut feel.

When a job comes in below that floor, you have three choices. Reprice it to clear the bar. Decline it. Or take it strategically — knowing you're breaking even or losing money — because it leads to something more valuable. That third option should be rare and deliberate, never accidental.

Job Costing in Real Time: Making It Practical

None of this works if it lives in a spreadsheet you look at once a quarter. Job costing needs to be part of how you quote every single job.

The simplest version is a job cost worksheet — a template with every cost category pre-built, where you plug in the job-specific numbers and it spits out your price floor and expected margin. You build it once and use it every time. In my company we use the Haulers' Edge AI tool to speed this up — upload photos, add job notes, get a price range and margin estimate back in minutes. Whether you use a purpose-built tool or a basic spreadsheet, the goal is the same: remove guesswork from your pricing.

For landscapers it's materials, labor hours, and disposal. For cleaners it's supplies, labor, drive time, and overhead allocation. For plumbers and HVAC it's parts, labor, and truck costs. The categories shift by trade. The discipline of tracking them all is universal.

Over time, job costing builds something even more valuable than accurate pricing — it builds a database of what your jobs actually cost. Patterns emerge. You start seeing which job types consistently hit your margin target and which ones consistently miss. That data shapes your entire business strategy: what to market, what to price more aggressively, and what to stop doing altogether.

What Happens When You Skip This

Businesses that don't cost their jobs properly don't usually fail dramatically. They fail slowly. Revenue looks okay. Busyness feels like progress. And then over months and years, thin margins compound into no reserves, no buffer, no ability to invest in growth. The owner is working harder and harder for less and less and can't figure out why.

The answer is almost always hiding in the gap between what they thought jobs cost and what they actually cost.

Close that gap and everything downstream gets easier — pricing, profitability, growth, and eventually the ability to build a business that doesn't need you on every job.

> *Know what every job costs. Price it accordingly. Do it every time.*

That's not complicated. It's just disciplined.

Go Deeper

For more on this and everything else in the book, the Haulers' Edge Newsletter goes deeper every week. Scan the code or subscribe free at HaulingHubb.com.

CH. 4 PLUG THE LEAKS

"A business without systems is just a job that follows you home." —Justin Hubbard

Back when I started Junk Pros in 2014 I was like a kid in a candy store. Said yes to everything. Junk removal, light hauling, full-house moves — if someone was willing to pay, we were willing to show up.

That lasted until we took on a big household move without the right equipment or experience. The job was a mess. It dragged on for hours, the customer wasn't happy, and it knocked our scheduling sideways for regular junk removal jobs the rest of the week. We strayed from what we were good at and paid for it in efficiency, reputation, and money.

The fix was obvious once I saw it. I called a local moving company and proposed a trade — I'd send them anyone who needed movers, they'd send me anyone who needed junk removed. We each focused on what we did best. The crews weren't dead on their feet by 4pm. Both businesses got more efficient and more profitable.

That partnership taught me something I use every single day: the businesses that win aren't the ones that do the most things. They're the ones that do the right things and cut everything else loose.

Know Your Winners

Every service business has a handful of services that carry the operation — jobs that come in consistently, deliver strong margins, and don't create chaos when you run them. Everything else is either a complement to those winners or a drag on them.

The question most owners never sit down to answer honestly is: which is which?

Start by listing every service you offer, including the ones you only say yes to occasionally. They all consume time and resources so they all belong on the list. Then rate each one on two dimensions — profitability and difficulty.

Profitability is margin. How much is left after the true cost of doing the job? If you haven't been tracking this, Chapter 3 gives you the framework. Difficulty is execution cost — does the service require special equipment, specific expertise, or create a disproportionate number of headaches, complaints, or delays?

Plot them on a simple two-by-two matrix. High profit and low difficulty in one corner — those are your winners. High profit but high difficulty — worth keeping but needs streamlining or repricing. Low profit and low difficulty — marginal, maybe useful as add-ons. Low profit and high difficulty — cut them. They are consuming your best resources and returning the least. If your analysis shows every service landing in the bottom half, the problem isn't your service mix — it's your pricing. The matrix tells you what to cut; Chapter 5 tells you how to reprice what's left.

For us the answer was clean once we ran the exercise. Dumpster rentals, our Grizzly Bag pickup service, and standard junk removal landed in the winner quadrant every time. Hauling heavy specialty items like old boilers and safes — high difficulty, thin margins, tied up our best crew for hours. Gone. Full house moves — not our equipment, not our expertise. Referred out. Same thing with our light-landscaping division. The business got lighter, faster, and more profitable almost immediately.

The fox that chases two rabbits catches neither. Better to be excellent at three things than mediocre at ten.

Cut What Doesn't Belong

Once you've identified your winners, everything else needs to earn its place or get cut. This is where a lot of owners hesitate because cutting a service feels like losing revenue. It's not. It's recovering the time, crew capacity, and mental bandwidth you've been subsidizing with your best resources.

Ask yourself honestly — are there services that make you cringe when a customer asks for them? Jobs you take because you feel like you have to, not because they make sense? For me it was answering calls about piano moves and pool table disassembly. Every time, I knew before I quoted it that it was going to be a bad day.

For each service that isn't pulling its weight, run a quick evaluation. Does it clear your minimum margin threshold? We use 40% gross margin as our floor — anything below that needs to be repriced or dropped. Does it create opportunity cost? Every hour your crew spends on a low-margin job is an hour they're not available for a high-margin one. Does it generate complaints or delays that create reputation risk? A bad review tied to work that wasn't worth taking financially is the worst kind of loss.

When you decide to cut a service, do it cleanly. Finish existing commitments. Give regular customers advance notice. Frame it as a focus shift, not a failure. Update your website and listings so you stop fielding calls for work you no longer do. Keep a referral list so you can hand those customers off rather than just turning them away.

Cutting also applies to customers, not just services. Some clients consistently overfill dumpsters, dispute invoices, demand discounts, or generate more management time than the job is worth. Set minimum charges. Define your service area firmly. And when a client is consistently more trouble than the margin justifies, walk away. The principle is the same everywhere — remove what drains you so the profitable parts of your business can breathe.

Build Partnerships for the Gaps

Cutting a service doesn't mean leaving customers stranded. The right move is building a network of partners who handle what you don't — so you can confidently say "we don't do that, but I know someone great who does" and mean it.

Look for businesses that naturally connect with yours. In hauling and junk removal that might be movers, cleaners, recyclers, scrap dealers, donation centers, contractors, or restoration companies. In landscaping it might be irrigation specialists, fence companies, or tree services. In cleaning it might be organizers, painters, or carpet companies. The overlap points are where partnership opportunities live.

The simplest partnerships are referral-based — you send them overflow work, they send you theirs. No money changes hands, just mutual trust that each of you will take care of the customer. That kind of arrangement compounds over time. My moving company partner has sent me hundreds of junk removal jobs over the years. I've done the same for them.

Before committing to a partnership, vet it the same way you'd vet a hire. Do their quality standards match yours? A bad experience with your referral reflects back on you. Check their reviews, confirm they're properly licensed and insured, and start with a small test referral before sending them anything significant. A partnership with the wrong company is worse than no partnership at all.

Done right, your referral network becomes an extension of your business — a way to serve every customer even when the job isn't yours to take.

Route Smarter

With your service mix tightened, the next place to plug leaks is how your trucks move through the day.

I used to schedule jobs in the order they came in. One morning we'd be downtown, afternoon out in a suburb 40 minutes away, then back to the city by evening. The truck was running a marathon. Fuel costs piled up. The crew was windshield-time tired before noon. And we were completing fewer jobs than we should have been.

Routing is simple to fix and immediately profitable. Group jobs by geography. Assign specific zones to specific days. Plan the most efficient sequence each morning before the first truck leaves the yard. Include dump runs in the route so you're never stuck with a full load at 4pm trying to find an open transfer station.

Basic free mapping tools handle this for small operations. Purpose-built routing software can shave even more off at scale. Studies show smart routing reduces drive time and fuel consumption by 20–30% and meaningfully reduces vehicle wear *(American Transportation Research Institute, 2023)*. On a fleet running five days a week that's thousands of dollars a year recovered without adding a single job.

You also don't have to accept every time slot a customer throws at you. When booking, suggest windows that cluster with existing jobs in that area. Most customers are flexible if you're upfront about it. Running the schedule on your terms — not reacting to every individual request — is one of the quietest competitive advantages in this industry.

Own Your Backyard

When we started we thought a bigger service area meant more

business. Five counties had to be better than two, right?

It wasn't. Long drives burned fuel, exhausted the crew, put unnecessary miles on the trucks, and made same-day service nearly impossible. We eventually pulled back to our city and the towns immediately surrounding it. Revenue didn't drop. Efficiency jumped. We became the recognizable local company in those areas — our trucks were familiar, our name was known, referrals picked up because people saw us constantly.

A tighter service area means more jobs per gallon, less crew fatigue, faster response times, and a brand that feels local rather than generic. It's better to dominate a small territory than to be a forgettable option spread across dozens of zip codes.

To find your right footprint, map where your revenue actually comes from. Eighty percent of it almost certainly falls within a manageable radius. The outliers outside that radius — calculate what they actually cost to serve including drive time, fuel, and wear. Most of them aren't profitable when you do that math honestly.

Draw your boundary. Stop marketing outside it, or add a distance surcharge that makes those jobs worth the extra cost if you take them. Refocus your Google Ads, your SEO, and your word-of-mouth to your core area. Density beats distance every time.

Standardize Everything You Can

Look at your equipment. Multiple truck brands, several dumpster sizes, different tool systems across your crew. On the surface that feels like flexibility. In practice it's a hidden cost factory — more parts to stock, more training required, more scheduling complexity, more things that can go wrong in ways you haven't anticipated.

One of the highest-return decisions I made in my hauling business had nothing to do with marketing or pricing. When we built out the dumpster operation we made one choice that simplified everything downstream: one dumpster size, 20 cubic yards, period. Paired with one hook-lift truck style. That was it.

The operational effects were immediate. Our mechanic stocked parts for one truck type and got fast at working on it. Any crew member could jump into any truck without retraining. Any dumpster could go to any job. Scheduling became clean. Deliveries became predictable. But the financial effect is what made it a profit decision rather than just an operations decision — maintenance costs dropped, training time dropped, and operational errors dropped. Over a full year the recovered margin from that single standardization decision ran into the thousands. Not from one dramatic saving but from dozens of small inefficiencies that simply stopped occurring.

The same logic applies beyond trucks and equipment. Standardize your cordless tools to one battery platform so you're not buying and stocking six different charger types. Standardize how trucks get loaded so every crew member does it the same way every time. Standardize your job completion checklist so quality doesn't depend on who's running the job that day. Standardize how new hires are trained on day one versus day thirty so the learning curve is predictable and the output is consistent.

Every place you can replace variation with a repeatable standard is a place where quality goes up and cost goes down. Standardization isn't a limitation on what your business can do. It's the foundation that lets it do more — with less friction, less waste, and better margins than the operation running on improvisation.

The Discounts You Never Agreed To

There's a category of margin leak that never shows up as a line item because it never gets recorded. It's the extras your crew gives away on jobs without thinking — and without telling you.

The customer asks them to move a piece of furniture while they're there. They say sure. The job was quoted for a specific scope and it expands on-site because the crew didn't want to have the conversation about the overage. The customer asks if they can add a few extra items and the crew says no problem. Each instance is a few dollars. Across a month of jobs they're hundreds of dollars of unbilled work that your team delivered and your business absorbed.

This isn't a crew problem — it's a systems problem. Your crew hasn't been given clear authority over what's included and what isn't, so they default to yes because yes feels easier than the conversation.

Fix it with two things. A clear scope definition on every job before the crew arrives — what's included, what triggers an additional charge, and what the crew is authorized to add without calling. And a simple add-on process that makes it easy for the crew to capture extras rather than give them away. “That's outside the original scope but we can handle it for an additional fee” needs to be a sentence in your crew's vocabulary — trained, practiced, and expected.

The same principle applies to time. If a job takes significantly longer than quoted because of conditions that weren't visible at estimate, that overage needs to be captured. Document the reason, communicate it to the customer before absorbing it, and charge accordingly. A crew that routinely finishes jobs that take twice as long as estimated without flagging it is quietly erasing margin on every one of those jobs.

Track unbilled extras as a metric. If you're not capturing them, you don't know how much you're losing. The invisible discount is the most expensive one because you never see it leave.

Prepare for Peak Season Before It Arrives

Most service businesses have a peak season and most arrive at it underprepared. One week it's slow, the next the phones don't stop — and the scramble to catch up costs money, costs jobs, and creates operational chaos that a little preparation would have prevented entirely.

Peak season preparation happens in the slow season. Not during the rush.

Eight weeks before your market's busy period run through this checklist. Staffing — do you have the crew you need to handle peak volume or are you going to spend the first month of your busy season interviewing while jobs pile up? Equipment — are your vehicles serviced and ready or are you going to discover a problem when you're booking three weeks out? Pricing — have you adjusted your rates to reflect peak demand or are you going to run your busiest weeks at off-season prices because you forgot to update them? Marketing — are your ad budgets ready to scale with demand or are you going to run out of budget by noon every day during your peak weeks?

Each of these preparation failures costs real money. A truck that breaks down during peak season doesn't just cost the repair — it costs every job that truck would have run while it was down. An understaffed operation during peak weeks turns away jobs or delivers rushed service that generates complaints instead of reviews. A pricing structure that wasn't updated before demand spiked leaves margin on the table that competitors are capturing.

The slow season isn't downtime. It's the window where you build the foundation that makes peak season profitable instead of just busy. The operators who treat it that way show up ready to capture the rush. The ones who coast show up behind and spend three months catching up.

What to Do When It's Slow

Slow periods feel like a problem. They're actually an opportunity that most businesses waste by coasting.

The operators who compound their advantage fastest treat every slow period as a building window. When the phones aren't ringing you have something you don't have in peak season — time. The question is whether you spend it recovering or building.

Training is the highest-return use of slow season time. The team member who needs more reps on customer communication, the new hire who got thrown into peak season before they were fully ready, the process gaps that everyone knows exist but nobody had time to address — slow season is when you close those gaps. Run call practice sessions. Walk through your SOPs together. Review recorded customer interactions and debrief what went well and what didn't.

Systems work is the second highest return. The processes that never got documented because there was always a more urgent job. The CRM automations that got set up halfway. The routing analysis you've been meaning to run. The job costing review that would tell you which service lines are actually profitable. None of this happens during peak season because there's no time. Slow season is when you build the infrastructure that makes peak season more profitable.

Pipeline building is the third. Reach out to commercial contacts you've been meaning to follow up with. Run a reactivation campaign to past customers who haven't booked in twelve months. Post consistently on your Google Business Profile and social media to keep your visibility active during the months when competitors go quiet. The operators who market through slow seasons show up to peak with a warmer market than the ones who went dark for four months.

Slow is temporary. What you build during it isn't.

The Gutter Cleaning Lesson

In 2017 I started a gutter cleaning business. Most competitors tried to do everything — cleaning, repairs, installs, residential and commercial. Their operations were clunky and their sales process was slow. I secret-shopped several of them and the pattern was consistent: customers waited a week for an estimate and another few weeks to get on the schedule.

I went the opposite direction. One service — cleaning only. One customer type — small homes. Two prices — $127 for one story, $167 for two stories. No estimates, no site visits. I used Google Earth and Zillow to verify the house size before confirming the price. If it fit the profile, I quoted it on the call and booked it on the spot.

We booked 9 out of 10 callers. The crews needed one ladder type and basic supplies. Since all the homes were similar in size, we grouped them by neighborhood and knocked out more houses per day than any competitor could match. The model never changed. Year one was solid. Year two we tripled. Year three we grew again and sold the business to a national competitor for a six-figure exit.

Standardization isn't about limiting what you can do. It's about doing what you do so consistently and efficiently that the business becomes both highly profitable and highly sellable. Focus compounds.

Align Your Marketing With Your Operations

There's no point tightening your operations if your marketing is still promising services you've cut or geography you no longer cover. Marketing is the outward expression of your internal strategy — it should reflect exactly what you do, for whom, and where.

When we refined our focus at Grizzly Junk Pros, we updated everything. The tagline shifted from the generic "Junk Removal and Light Hauling" to something specific: "Fast, Friendly Junk Removal & Dumpster Rental — Serving Homeowners and Contractors." One line told people exactly what we do and who we serve. References to moving and other dropped services disappeared from the website completely.

The calls changed almost immediately. Instead of time-wasting inquiries about services we didn't offer, we started getting calls like "I saw you specialize in garage cleanouts — do you also take yard debris?" and "I need a 20-yard dumpster — I found you on Google in my town." Those were the exact leads we wanted.

Make your website work as a filter, not just a magnet. Call out what you don't do as clearly as what you do. "We don't remove hazardous materials" on your FAQ page eliminates those calls before they happen. "Pianos and safes by special arrangement only" sets the expectation upfront. Meanwhile, lead with the jobs you want. If hoarder cleanouts are a profitable niche for you, say so explicitly — "We handle whole-home hoarder cleanouts, priced by project after assessment." That repels the bargain hunters and attracts the serious customers who need the work done right.

Tighten your ad targeting to match your service area. If your Google Ads are running across three counties but you only want jobs in one, you're paying for clicks you can't profitably serve. Lock the geography. Watch your cost per lead drop and your close rate climb.

Say No Like You Mean It

Warren Buffett said the difference between successful people and really successful people is that really successful people say no to almost everything. You can't run that extreme in a service business — but the spirit of it is exactly right.

Not all revenue is good revenue. Some jobs cost more in time, crew capacity, and risk than they return in margin. The sooner you internalize that, the more profitable your business gets.

Say no when the job is outside your core scope and the risk is high — a piano in a basement, a boiler in a crawlspace, hazmat you're not licensed to handle. Say no when the customer is negotiating against your real costs. Say no when the location makes the job unprofitable before a single item gets moved. Say no when your gut tells you this one is going to be a problem — your instincts have usually already run the math.

When you do say no, do it cleanly. Thank them, explain briefly, and hand them to someone who can help. "That's not something we handle, but I know a company that's great at it — let me give you their number." That response builds more goodwill than a reluctant yes that ends in a bad job.

Your crew needs to be empowered to do the same. If a customer asks them to demo a shed mid-cleanout that wasn't in the quote, they shouldn't absorb that work to avoid an awkward conversation. Train them to pause and call it in: "We're happy to do that — let me get you a price for the additional work." Scope creep absorbed silently is margin lost permanently.

Charge What Hard Jobs Actually Cost

Some jobs are worth taking — but only at a price that makes them worth your while. We built the PITA factor directly into our quoting process. Pain In The Ass factor. If a job carries extra difficulty, unusual disposal costs, or elevated risk, the price reflects it without apology.

A standard truck of household junk might run $795. That same load if it's primarily construction debris mixed with concrete runs $1,250 or more — it's heavier, costs more to dump, and works the crew harder. A third-floor cleanout with no elevator gets a labor surcharge. Every job like that has a premium built in.

We also use what I call the walk-away price — a number high enough that if the customer says no, we're relieved, but if they say yes, we're well compensated for the headache. You'll find that the serious customers who actually need difficult work done will often say yes to a fair premium price when you explain what's driving it. The ones who push back hard were going to be problems anyway.

Give customers options when the job has a high-cost variable. Full house cleanout including the boiler removal for $2,000, or everything except the boiler for $1,500 with a referral to a specialist. That transparency builds trust and lets the customer make an informed decision rather than feeling like they're just absorbing a number.

And when customers push for discounts based on what a competitor quoted — hold firm. "We might not be the cheapest option, but our pricing reflects what it actually costs to do this job right with the proper insurances and pay our crew fairly." The customers who respect that are the ones worth keeping. The ones who walk were going to be a problem at every invoice anyway.

> *Your business grows just as much from what you decline as from what you accept.*

Say no strategically, price hard jobs honestly, and watch the quality of your work — and your margins — go up at the same time.

Go Deeper

For more on this and everything else in the book, the Haulers' Edge Newsletter goes deeper every week. Scan the code or subscribe free at HaulingHubb.com.

CH. 5 STOP GUESSING WHAT TO CHARGE

"You don't get paid for the hour. You get paid for the value you bring to that hour." —Jim Rohn

We quoted a full-house cleanout at $600 because that's what it had always been. After running the actual job cost math from Chapter 3, the real cost on that job was $540. We were making $60 on a job that took six hours and two crew members. That's $5 an hour of profit per person. We were busy. We were not making money.

That's what pricing from memory produces. Not a disaster you can see — a slow bleed you can't find until you actually run the numbers.

This chapter gives you the framework to price from math, position from confidence, and hold your number when a customer pushes back.

Know What the Market Is Paying

Before you set a single price, you need to know the playing field. Market benchmarks tell you what customers expect to pay in your region and what your competitors are charging. Without that context you're pricing blind.

For junk removal and hauling, here's the general landscape at time of writing. A full truckload of residential junk runs roughly $400–$800 depending on market, with urban coastal markets at the high end and rural markets at the low end. Major national franchises like 1-800-GOT-JUNK charge a premium for brand recognition — a full truck in a major city can exceed $800. Single large item pickups run $100–$180 with most established companies. Roll-off dumpster rentals average around $300–$600 per week for 10–20 yard containers, with the national average near $400 *(HomeAdvisor, 2024)*. Small demolition work runs $2–$8 per square foot for interior demo, with simple shed removals averaging around $650 and ranging up to $3,000 depending on size and complexity. Prices in all categories trend upward with inflation — verify current market rates in your area before setting your own benchmarks.

For a landscaper, house cleaner, or HVAC tech, the benchmarks look different but the research process is the same. Check competitor websites. Call two or three as a customer and get a quote for a standard job. Search "[your service] pricing [your city]" and review what comes up on aggregators like Angi or HomeAdvisor. Build a simple list. Know the floor, the ceiling, and where most of the market sits.

This research does two things. It tells you whether you're leaving money on the table. And it tells you where you can justify a premium if your service is genuinely better than what's out there.

Never price in isolation. The market is your first reference point.

Build Your Price From the Bottom Up

Knowing the market tells you what customers will pay. Knowing your costs tells you what you need to charge. Those two numbers together define your pricing window — the range where you're both competitive and profitable.

We covered job costing in depth in Chapter 3. Here's how it connects to pricing.

Every job has a true cost that includes direct labor with payroll burden, your own time at its real value, fuel for the full route, disposal fees by material type, materials and supplies, equipment depreciation, and an overhead allocation that carries each job's share of your fixed costs. Add those up and you have your cost floor — the absolute minimum below which every job loses money.

On top of that floor you add your margin. For service businesses, a healthy gross margin on individual jobs runs 40–60% depending on the service type, complexity, and your market position *(IBISWorld, 2023)*. The formula is straightforward:

Price = Job Cost ÷ (1 − Target Margin)

So if a job costs $170 and you're targeting a 40% gross margin: $170 ÷ 0.6 = $283. Round to $285 or $297 depending on your charm pricing preference and you have a floor-based price you can defend.

Now sanity-check it against the market. If your cost-based price is $283 and competitors are charging $350–$450 for the same job, you have room to price higher and improve your margin. If your cost-based price is $283 and competitors are charging $200, you either have a cost problem to fix or a value story to build — but you don't drop below your floor. Ever.

Set a minimum charge and hold it. No job leaves your yard for less than a number that covers your truck roll, your labor, and your disposal. In junk removal that minimum typically runs $75–$150 depending on market. National franchises hold $129 as a standard floor. Your minimum is your break-even insurance — it prevents you from sending a crew across town for a bag of trash that earns you nothing.

Where to Position Yourself

Once you know your cost floor and the market range, you need to decide where you want to sit within that range. This is a strategic choice, not a default.

The budget position — sitting at the low end of the market — wins price-sensitive customers but attracts bargain hunters who will leave you for anyone $5 cheaper. Margins are thin, volume pressure is constant, and you're one cost increase away from losing money on jobs. It's a hard place to build a sustainable business.

The premium position — sitting at or above the market midpoint — requires a genuine reason why. Better response time. Cleaner presentation. Stronger reviews. Guaranteed pricing with no surprise fees. A more thorough job. If you can articulate a real difference, you can hold a higher price and attract customers who value quality over cheapness. 1-800-GOT-JUNK is almost never the cheapest option in any market. They're consistently one of the most booked because brand, professionalism, and reliability command a premium that a segment of customers will always pay.

My company is not close to the cheapest option in our market. That's a deliberate choice. We compete on reliability, presentation, and the confidence that comes with a track record. The customers we lose to cheaper competitors are usually the ones who would have been problems anyway.

Competing purely on price is a race to the bottom and there's always someone willing to lose money faster than you. Compete on value instead.

The Customers You Scare Off Were Never Going to Be Good Customers

The most common reason service operators don't publish their pricing is fear. Fear that a competitor will undercut them. Fear that customers will sticker-shock and disappear. Fear that showing the number takes away negotiating leverage.

Here's what actually happens when you publish honest pricing on your website.

The customers who would have called just to get a price and then shopped it against three competitors — they self-select out before they ever call you. You don't spend fifteen minutes on a call that was never going to close. You don't send an estimate that was always going to be used as leverage against a cheaper option. Those customers remove themselves from your pipeline before wasting your time.

The customers who call after seeing your pricing are different. They've already seen the number. They're still calling. That means they're either comfortable with the price, they value something about your operation beyond the lowest rate, or they have a question that the pricing page didn't fully answer. Any of those three is a productive conversation. You're not starting from zero on every call — you're starting from a customer who already knows what they're getting into.

Honest pricing on your website also makes you findable in AI search. When someone asks an AI assistant what junk removal costs in your city, the AI pulls from pages that answer that question directly. If your pricing page gives a clear, honest answer and your competitor's says “call for a quote” — you're in the AI answer and they're not.

Publish your pricing. Not a vague range that tells the customer nothing. A real explanation of what affects the cost, what's included, and what the typical job runs. The customers you lose because of it were always going to be a problem. The ones who stay are the ones you want.

Use Psychology to Make Your Price Land Better

The math sets your price. Psychology determines how customers hear it.

Anchoring works because people evaluate prices relative to the first number they encounter. If you present a premium option first — a full-service whole-home cleanout at $2,500 — your standard single-room option at $650 feels reasonable by comparison. Show the high end first. Always.

Charm pricing — ending prices in 7s or 9s rather than round numbers — exploits left-digit bias. $297 registers as meaningfully less than $300 even though the difference is three dollars. It works better in consumer-facing marketing like ads and websites than in direct conversations with commercial clients, where round numbers read as more professional.

Bundling increases ticket size while making customers feel they're getting a deal. A cleanout plus haul-away plus dumpster drop as a single package price — slightly less than the sum of parts — gets a customer to say yes to more in one transaction. It also differentiates you from competitors who don't offer combined services.

Tiered options let you serve different budgets without discounting your core offer. Curbside pickup at a lower rate for the customer who wants to do the carrying. Full-service interior removal at the standard rate. White-glove handling for specialty or high-volume jobs at the premium rate. The customer self-selects. You protect your margin at every level.

Decoy pricing nudges customers toward the option you want them to choose. If you offer three tiers priced at $150, $300, and $350, the $300 option starts to look like a poor value relative to the $350 option — which makes most customers choose the large. The middle tier exists to make the top tier look like a deal.

None of these replace honest math and solid positioning. But they shape how customers experience your prices, and that matters at the moment of decision.

How to Hold Your Price When Pushed

You've set a fair price built on real costs and market research.

Now a customer says it's too high.

This is where most operators lose money they never had to lose. They panic, discount immediately, and train their customers that the first number is never the real number.

Don't do that.

When a customer says "that seems expensive" the right response is not to justify every line item or drop the price. State the number confidently. Pause. Let them process. Then if they object, walk them through what's included — not what it costs you, but what they get. Two crew members handling all the heavy lifting. Full disposal included with no hidden fees. The space cleared, swept, and finished in a single visit. Insurance and licensing that protects them if anything goes wrong.

Frame the comparison they're actually making. A dumpster rental for a DIY approach might cost them $400–$600 plus their own time, labor, and dump fees if they go over weight. Your full-service rate is competitive with that once you factor in the work they're not doing.

When they cite a cheaper competitor, respond without attacking. "That could be a good option — just confirm what's included and whether they're fully insured. We've seen low quotes that didn't cover disposal or left the customer with a bill at the end for overages. Our price is all-in with no surprises." You've planted a legitimate question without saying anything negative about the competition.

If you're going to offer something extra to close a difficult job, make it strategic — a small add-on that costs you little but feels valuable to them. Not a percentage off your actual price. A discount trains them to negotiate every time. A bonus gesture closes the deal without eroding your rate.

And know when to walk away. A customer who only buys on price is not a customer you want to build your business around. They'll leave you for anyone cheaper, dispute invoices, and create more management friction than the margin justifies. Let them go. The time you recover is worth more than the revenue you'd generate at a margin that doesn't sustain your business.

Price Hard Jobs What They're Actually Worth

Chapter 4 covers the PITA factor and walk-away pricing in detail — the premium you build into quotes for jobs that carry extra difficulty, risk, or cost, and the minimum price at which you'd be relieved if the customer said no. Both apply directly here: non-standard jobs need non-standard prices, and giving customers options on complex work lets them make a real decision rather than react to a number they don't understand.

Raising Prices Without Losing Customers

Your costs go up every year. Fuel, disposal fees, labor, insurance — the floor rises constantly. Your prices need to rise with it or your margins will erode.

The key to raising prices without drama is frequency and communication. Small, regular increases are almost always absorbed better than large infrequent ones. A 5% increase annually registers as a routine cost adjustment. A 20% increase after three years of holding flat triggers sticker shock and customer calls.

When you raise prices, tell your regular customers before it happens and tell them why. "Starting March 1 our rates are increasing approximately 5%. Disposal and fuel costs in our area have risen significantly and we've held off as long as we could. We remain committed to the same level of service and appreciate your continued business." That message, sent a few weeks in advance, almost never generates pushback from customers who value you. They know costs go up. They respect the transparency.

For price-sensitive long-term customers, consider a smaller increase than new customers see — not as a permanent policy, but as a one-time acknowledgment of the relationship. It costs you almost nothing and generates significant goodwill.

One tactic worth using carefully: adjusting the service spec rather than the headline price. If your dumpster rental was $400 for 7 days with 2 tons included and disposal costs have risen, you might keep the $400 rate but adjust the included weight to 1.5 tons. The headline price stays the same. The cost structure improves. Additional weight becomes an add-on that customers only pay if they need it. We've used this successfully at Grizzly Junk Pros and it's a cleaner adjustment than straight price increases in competitive markets.

What you should never do is cut service quality to offset rising costs while holding prices. You'll protect your margin for one quarter and destroy your reputation for the next three years. If something has to give, be transparent about it and frame it as a trade the customer is choosing, not something you're doing without telling them.

> *Price is a signal. What yours signals about your business is a choice.*

Charge what your work is worth, defend it confidently, and adjust it honestly as your costs change.

Go Deeper

For more on this and everything else in the book, the Haulers' Edge Newsletter goes deeper every week. Scan the code or sub-

scribe free at HaulingHubb.com.

CH. 6 THE TAX TRAP MOST SERVICE BUSINESSES WALK RIGHT INTO

> *"In this world, nothing is certain except death and taxes." — Benjamin Franklin*

The tax trap doesn't announce itself. It doesn't show up as one catastrophic mistake. It's a slow bleed — the wrong business structure costing you thousands a year, the deductions you never claimed because nobody told you about them, the quarterly payments you skipped because cash was tight, and then the April bill that hits like a truck and wipes out everything you built over the previous twelve months.

I've watched solid operators get blindsided by taxes not because they were careless but because they were never taught how the system works. And here's the thing — the tax code is not designed to hurt business owners. It's actually written with significant advantages built in for people who own businesses and real assets. The employees paying taxes through a W2 get almost none of those advantages.

This chapter isn't about dodging anything. It's about understanding the system well enough to stop leaving money on the table every single year. If you already have an LLC, an S-Corp election, and a CPA handling your quarterly estimates, the first two sections are review — skip to the Section 179 and depreciation section. That's where even operators who think they've handled taxes are still leaving money behind.

The First Mistake: Wrong Business Structure

Most service businesses start as sole proprietorships because it requires nothing — you're in business the moment you start working. No paperwork, no filing, no setup cost. And for the first few months, that's fine.

The problem is what sole proprietorship costs you as revenue grows.

As a sole proprietor, every dollar of net profit is subject to self-employment tax — which is 15.3% on top of your regular income tax *(IRS, 2024)*. That 15.3% covers both the employer and employee share of Social Security and Medicare. As an employee, your employer pays half of that. As a sole proprietor, you pay all of it. On $100,000 of net profit that's $15,300 in self-employment tax before you've paid a single dollar of income tax. On top of that, your personal assets — your house, your savings, your personal bank account — are fully exposed to any business liability. One lawsuit, one serious accident, one judgment against your business and everything you own is potentially on the table.

The LLC fixes the liability problem. A single-member LLC creates a legal separation between your business and your personal assets. It doesn't automatically change your tax situation — a single-member LLC is still taxed as a sole proprietorship by default — but it protects everything you've built personally from business risk. For a service business operating trucks, dealing with customers' property, and employing people, that protection is non-negotiable. An LLC costs a few hundred dollars to set up and a small annual fee to maintain depending on your state. It is the minimum structure every service business should be operating under.

The S-Corp Election: Where Real Tax Savings Live

Once your business is generating consistent net profit above roughly $50,000–$60,000 a year, the S-Corp election deserves a serious look. This is where the tax code hands service business owners a genuine advantage that most never use.

Here's how it works. When you elect S-Corp status, your business becomes a pass-through entity that allows you to split your income into two buckets — a reasonable salary and owner distributions. You pay payroll taxes only on the salary portion. The distributions are not subject to self-employment tax.

Example. Your business generates $150,000 in net profit. As a sole proprietor or standard LLC, you pay 15.3% self-employment tax on the full $150,000 — that's $22,950 just in SE tax before income tax. With an S-Corp election, you pay yourself a reasonable salary of $60,000 — which is subject to payroll taxes — and take $90,000 as a distribution, which is not. Your SE tax applies only to the $60,000. That's roughly $9,180 in SE tax instead of $22,950 — a difference of over $13,000 in a single year *(IRS, 2024)*. The savings compound every year you operate at that level.

The S-Corp does come with added complexity. You need to run actual payroll, file additional tax returns, and pay yourself a salary that the IRS considers reasonable for someone doing your role. If you pay yourself $1 to avoid payroll taxes the IRS will recharacterize the distributions and hit you with penalties. The salary needs to be defensible — typically somewhere around what you'd pay someone to do your job. But the compliance cost of the S-Corp — an accountant and a payroll service — is almost always a fraction of the tax savings at the revenue levels where it makes sense.

The general rule of thumb most CPAs use: if your business is netting $50,000 or more consistently, have the S-Corp conversation. If you're netting $80,000 or more, the savings almost certainly justify the added cost and paperwork.

Quarterly Taxes: The Bill Most Operators Ignore Until It's Too Late

When you work a W2 job, taxes come out of every paycheck automatically. You never see the money so you never miss it. As a business owner, nobody withholds anything. The money lands in your account and it feels like yours.

It's not all yours.

The IRS expects self-employed business owners to pay estimated taxes quarterly — in April, June, September, and January. If you don't, you get hit with underpayment penalties on top of the full tax bill when you file. And if you've been spending the money as it came in without setting anything aside, April becomes a crisis.

The fix is simple but requires discipline from day one. Set aside 25–30% of every dollar of net profit into a dedicated tax account and treat it as untouchable. Not 20%. Not "I'll figure it out later." Twenty-five to thirty percent, set aside automatically, every time money comes in. When quarterly payments are due, you make them from that account. When you file in April, the money is already sitting there.

If you're operating as an S-Corp with payroll, your salary withholdings handle a portion of this automatically. But you still owe tax on the distribution income, so the reserve habit matters regardless of structure.

The specific quarterly amounts depend on your prior year tax liability and projected current year income — your CPA calculates the right figures. What matters operationally is that the habit exists so the money is always there when the payment is due.

What You Can Deduct: The List Most Operators Leave Money On

Every legitimate business expense reduces your taxable profit dollar for dollar. Most service business operators claim the obvious ones — fuel, labor, equipment — and miss a significant portion of what they're legally entitled to deduct.

Here's what belongs on your deduction list that often gets missed.

Vehicle expenses. You can deduct actual vehicle operating costs — fuel, insurance, maintenance, repairs, registration — or use the IRS standard mileage rate (check IRS.gov for the current year's rate — it adjusts annually). Track every business mile. For a service business running trucks all day this adds up to significant money. If you use a personal vehicle for any business purpose — driving to a job site, picking up supplies, meeting a vendor — those miles count.

Home office deduction. If you manage your business from a dedicated space in your home — even a corner of a room used exclusively for business — a portion of your rent or mortgage, utilities, and internet is deductible. The space must be used regularly and exclusively for business. The simplified method allows $5 per square foot up to 300 square feet. The actual expense method calculates the real percentage of your home used for business and applies it to your actual costs. For most small operators the home office deduction is several hundred to several thousand dollars a year that never gets claimed.

Phone and internet. The percentage of your phone bill and internet service used for business is deductible. For most service businesses running a service business from their phone that percentage is high.

Tools, equipment, and supplies. Everything you buy to do the work — tools, safety gear, tarps, straps, supplies — is deductible. Keep receipts. Track purchases. Don't let small supply runs go undocumented because they feel too minor to matter. Across a year they add up.

Marketing and advertising. Every dollar spent on Google Ads, website hosting, business cards, yard signs, uniforms with your logo, sponsorships, and any other marketing activity is deductible.

Professional services. Your accountant's fees, legal fees, business coaching, consulting — all deductible. The fee you pay your CPA to file your taxes is itself a tax deduction.

Education and training. Books, courses, conferences, and subscriptions directly related to running your business are deductible. This book qualifies if your reader buys it for their business.

Meals with business purpose. Business meals are 50% deductible when there's a legitimate business discussion. Keep notes on who you met with and what you discussed.

Insurance premiums. Business liability insurance, workers' compensation, commercial auto insurance — all deductible operating expenses.

The documentation habit is what makes deductions defensible. Keep receipts. Use a business credit card for every business expense so you have a clean transaction record. Don't run personal expenses through the business and don't run business expenses through personal accounts. That separation protects you in an audit and makes your books accurate.

Section 179 and Bonus Depreciation: The Equipment Write-Off Most Operators Underuse

When you buy equipment — trucks, trailers, dumpsters, tools — the IRS normally requires you to depreciate that cost over several years rather than deducting it all at once. Section 179 is the exception that lets you deduct the full purchase price of qualifying equipment in the year you buy it, up to the annual limit.

The Section 179 deduction limit adjusts annually — check the current year's limit at IRS.gov, but it's typically well over a million dollars. For a service business buying trucks and equipment, this is a powerful tool. If your business shows $100,000 in taxable profit and you purchase a $100,000 truck, Section 179 can bring your taxable income to zero for the year. At a 25% effective tax rate, that's $25,000 in taxes you don't pay.

Bonus depreciation works similarly — it allows accelerated deduction of asset costs in the year of purchase. The percentages and rules around bonus depreciation have been phasing down since 2023, so verify current rates with your CPA before planning around it.

Two important caveats. Section 179 only reduces your tax bill — it can't create a loss that carries back to prior years in most cases. And state tax rules don't always mirror federal Section 179 treatment, so your federal and state tax bills can look very different on a major equipment purchase year. Always confirm with your tax advisor before making a large purchase specifically for the tax benefit.

The strategic application: if you're having a strong profit year and you need a truck anyway, the timing of that purchase matters. Buying in December of a profitable year versus January of the next year can mean the difference between a significant tax bill and a near-zero one.

The Audit Risk Nobody Talks About

The IRS audits a small percentage of returns each year, but certain patterns draw attention. Knowing what those patterns are lets you stay clean.

Consistently reporting losses — especially if your business shows a loss in multiple consecutive years — can trigger scrutiny. The IRS has hobby loss rules that reclassify a business as a hobby if it doesn't show a profit in at least three of the last five years, which eliminates most of your deductions. Run a real business with real records and this isn't a concern. But if you're structuring your books to always show a loss to avoid taxes, you're building a problem.

Unusually high deductions relative to income attract attention. If your business makes $80,000 and you're claiming $75,000 in deductions, the math invites questions. That doesn't mean you can't have high deductions — it means every one of them needs documentation.

Misclassifying employees as independent contractors is one of the most common and expensive mistakes in service businesses. If you're directing when someone works, how they work, and what they work on, the IRS considers them an employee regardless of what your contract says. Misclassification carries back taxes, penalties, and interest. If you have regular workers who function as employees, structure them correctly *(IRS, 2024)*.

Cash businesses get extra scrutiny. If your revenue is primarily cash-based and your reported income is inconsistent with your lifestyle or apparent business activity, that inconsistency draws attention. Report accurately. Keep records. The risk of underreporting cash income isn't worth the exposure.

Sales Tax: The Variable Nobody Warns You About

Most service business owners assume services aren't taxable. In many states that's true. In others it's not, and the rules are specific enough that assuming wrong costs you real money.

Sales tax on services varies by state and sometimes by service type within the same state. In some states junk removal is fully taxable. In others it's exempt. In others still, the labor is exempt but disposal fees aren't. Landscaping, cleaning, and HVAC services face the same patchwork — taxable in one state, exempt in the next, partially taxable in a third.

The practical risk is real. If your state taxes your services and you're not collecting and remitting sales tax, you're personally liable for what should have been collected — potentially going back years, plus penalties and interest. The burden of proof falls on you, not the customer you forgot to charge.

Before you assume you're exempt, verify it. Your state's Department of Revenue website is the starting point. Your CPA can confirm the answer for your specific services in your specific state. This is a ten-minute conversation that could save you from a multi-year audit liability. Have it before you need it.

Find a Good CPA and Actually Use Them

Everything in this chapter is general education. The specifics of your situation — your entity structure, your state's tax rules, the right quarterly payment amounts, which deductions apply to your specific operations — require a professional who knows your books.

A good CPA who works with small service businesses is not an expense. They are an investment that typically returns several times their fee in identified savings, avoided mistakes, and legitimate deductions you wouldn't have found on your own. The IRS found that taxpayers who used a professional preparer received larger refunds and made fewer errors than those who filed independently *(IRS Taxpayer Advocate Report, 2022)*.

When you're evaluating a CPA, ask specifically whether they have experience with service businesses, whether they're familiar with S-Corp elections and the payroll requirements, and whether they do proactive tax planning throughout the year or just file returns in April. You want the proactive one. Filing a return is a look backward. Planning is how you actually reduce what you owe.

Meet with your CPA at minimum twice a year — once mid-year to review your current position and project your year-end liability, and once before year end to make any moves that affect the current tax year. December is when the decisions get made. April is too late to do anything but write the check.

The tax code was written with business and property owners in mind. It rewards people who create jobs, invest in assets, and build real enterprises. That description is you. Use what the code offers.

The operator who set up their S-Corp at $80K in revenue, started paying quarterly, and had one real planning conversation with a CPA in November saved $15,000 in year one — not through anything complicated, just by having the conversation this chapter told them to have. Your tax situation isn't unique enough that you can't do the same. The code is there. The savings are real. The only step between where you are and keeping more of what you earn is making the call.

> *Every dollar you keep is a dollar you earned twice — once when you made it, once when you didn't give it away unnecessarily.*

Go Deeper

For more on this and everything else in the book, the Haulers' Edge Newsletter goes deeper every week. Scan the code or subscribe free at HaulingHubb.com.

PART 2: MARKET POSITIONING

CH. 7 SECRET SHOPPING

"Know your enemy and know yourself and you can fight a hundred battles without disaster." — Sun Tzu

Most service business owners are so focused on their own operation they never look up to see what's happening around them. New competitor moves into the market. Existing one silently drops prices or adds a service. Customer expectations shift because someone else raised the bar. And you find out about it six months later when the phone gets quieter and you can't figure out why.

Competitive intelligence isn't paranoia. It's basic situational awareness. And secret shopping — experiencing your competition the same way a customer does — is the fastest, most honest way to get it.

Why This Matters More Than Most Operators Think

You don't know what you're competing against until you look. That sounds obvious but most service business owners never do it — or they do it once when they launch and never again.

The market moves. A competitor who was mediocre two years ago might have tightened up their operation, hired better people, and started running ads. A new entrant might be undercutting everyone on price to buy market share. A franchise might have moved into your territory with national brand recognition and a bigger marketing budget than your whole revenue.

None of that shows up in your own numbers until the damage is already done. Competitive research is how you catch it early.

The specific things you're looking for: what they charge, how they sell, what customers love about them, what customers hate about them, where they're weak, and what they're doing in marketing that's working. All of that is available to you without a single unethical move — it's just observation, research, and a phone call where you act like a customer. Which is exactly what a customer would do.

Done consistently, competitive research sharpens your pricing, improves your service, generates marketing ideas, and validates your strengths. It's one of the highest-return activities in the business and one of the most neglected.

When to Do It

Before you launch or add a service — non-negotiable. You need to know what you're entering before you set prices or build a marketing strategy.

On a regular schedule after that. A quick monthly scan of competitor websites and reviews. A deeper dive — including actual secret shopping calls — quarterly. Markets shift faster than most owners track.

When something feels off. Revenue is softer than it should be. The phone isn't ringing at the rate it was. Your close rate dropped. Those are signals that something in the competitive landscape changed. Investigate immediately rather than waiting for the next scheduled review.

Before major decisions. Raising prices, adding a service, investing in new equipment — check the landscape first. Someone else may have already tried what you're considering. Their reviews and results will tell you more than any theory.

How to Secret Shop Without Getting Caught or Getting Weird About It

Secret shopping is just calling a competitor as a customer would. Nothing elaborate required.

Don't use your main business number or a number they might recognize. Your personal cell is fine. If you've met the owner or they know your name, have a friend or family member make the call instead — give them a short script and what to listen for. In small markets where everyone knows everyone, this is worth the extra step.

Prepare a realistic scenario before you call. Pick a job type you handle regularly — a garage cleanout, a dumpster for a renovation, a demolition job. Know what you want to find out: their price, how they quote it, what's included, how they handle the call, what their availability looks like, whether they push back on price or crumble immediately.

When you call, pay attention to everything. How fast did they answer? Was the person confident or fumbling? Did they try to understand your job or just throw a number at you? Did they mention anything that could be a selling point you're not using? Did they ask for the booking or let you off the phone without trying to close?

The content of the call matters. The quality of the sales interaction matters just as much.

Write it down immediately after. Memory is unreliable — capture price, key phrases they used, how they handled the close, and your overall impression of whether you'd hire them if you were a real customer.

In my company we secret shop every major competitor in our market twice a year — all within the same week so we get a true real-time snapshot. Prices, response time, how they talk to customers. If someone has raised prices or added a new service, we know within days.

And yes — they're probably shopping you too. Take that as a

compliment.

One limit worth noting. Don't book an actual job and cancel just to see how they operate. That wastes their time and money. A phone call is almost always enough to get what you need.

Mine Their Online Presence

You can gather a significant amount of competitive intelligence without ever picking up the phone.

Their website is their pitch. Read it like a customer who knows nothing about them. What services do they lead with? Do they list prices or keep them hidden? What's their positioning — budget option, premium service, eco-friendly, local specialist? What do they not mention that you'd expect? A website that buries pricing often means they're higher and selling value over the phone. A website that leads with price usually means that's their primary competitive angle.

Reviews are the most honest data source you have on a competitor. Not their marketing — what their actual customers said after the job was done. Read the five-star reviews to understand what they do well. Read the one and two-star reviews even more carefully. Patterns in the complaints are your opportunity. If three different customers mention they showed up late, responsiveness is your competitive advantage to emphasize. If people consistently mention surprise fees, your all-in transparent pricing is a story to tell. Your competitors' negative reviews are your marketing brief.

Social media shows you what they're currently promoting and how engaged their audience is. Are they running seasonal specials? Posting before-and-afters that get shared? Building a community following or just broadcasting into the void? Look at what lands and what doesn't. Note which platforms they're active on and which they've abandoned — an abandoned channel is an opportunity for you to own that space.

Google search tells you where they rank for your shared keywords and whether they're investing in paid ads. Search your core service terms in your market. Who appears at the top organically? Who's running ads? Click their ad to see their landing page and offer. The Facebook Ads Library lets you search any business page and see what ads they're currently running —pricing, offers, messaging. All free intelligence, fully public.

General search on their business name occasionally surfaces news coverage, BBB listings, community mentions, or forum discussions you wouldn't find otherwise. Takes ten minutes and occasionally turns up something significant.

Build a Competitor Profile for Each One

Raw intel is only useful if it's organized. For each major competitor, build a one-page profile you update as you gather new information. Not a complex document — just a consistent record that gives you a snapshot at any moment.

The profile should capture their service lineup, pricing structure, branding and positioning, review rating with key themes from the feedback, their operational strengths, their weaknesses and complaint patterns, and any notable recent moves — new services, price changes, new marketing angles, fleet expansion.

At the bottom of each profile: your specific action items in response. What does what you learned about this competitor tell you to do, emphasize, or stop doing in your own business?

These profiles become working documents you update every cycle. After two or three rounds of research you'll have a clear picture of how each competitor has evolved and where the openings in your market are.

Build a Comparison Matrix

Beyond individual profiles, a side-by-side comparison matrix lets you see the whole competitive landscape at once.

Columns are your business and each major competitor. Rows are the factors that matter: services offered, price range, notable fees, unique value proposition, review rating, primary strengths, primary weaknesses, and marketing focus.

When you lay it out this way, gaps become obvious. Maybe nobody in your market offers same-day service and customers keep asking for it. Maybe you're the only one with online booking and nobody knows it. Maybe one competitor has expanded their service area into your core territory. Maybe you're the most expensive option but have the strongest reviews — which means you have a value story to tell more aggressively.

The matrix is also a useful tool for training. When a new hire asks "who else do customers call," you hand them the matrix instead of trying to explain the landscape verbally.

Update it twice a year minimum. More if the market is moving fast.

Turn Rivals into Referral Partners

Here's the counterintuitive part of competitive research. Some of your competitors are worth knowing personally, not just monitoring from a distance.

The hauling and service industry is smaller than it looks. The same owner you're competing against for a job on Tuesday might be the person who refers overflow work to you on Thursday when they're booked out. The competitor who doesn't offer dumpster rentals needs someone to send those customers to. The one who stopped doing demolition has clients asking for it every week.

I've built referral relationships with competitors that have sent me consistent business for years. The arrangement is simple — we each focus on what we do best, we refer what we can't or don't want to handle, and both operations run cleaner as a result. No contracts, no commissions in most cases. Just mutual trust and the understanding that taking care of each other's referrals reflects on the person who sent them.

To start, keep it low-key. If you cross paths at a transfer station or a Chamber of Commerce event, introduce yourself. Most owners in this industry are regular people who appreciate a genuine conversation over the assumption that every competitor is an enemy. You're not trading secrets — you're acknowledging that the market is big enough for more than one good operator and there's value in knowing each other.

Keep boundaries clear. You can be friendly without being naive. Don't share your pricing strategy, your client list, or anything operationally sensitive. And don't do anything that even approaches price-fixing or market allocation — that's not competitive camaraderie, that's a legal problem.

The competitors worth knowing are the ones who have standards that match yours. A referral from you is a reflection on you — send customers to people who will make you look good for sending them.

What to Do With Everything You Find

Intelligence that doesn't change behavior is just trivia. After every research cycle, identify two or three specific actions.

Maybe you discover a competitor is getting traction by explicitly advertising same-day service and you've never highlighted that you offer it. One sentence added to your Google Business Profile. Maybe their reviews consistently mention they sweep up after a job and yours don't — a two-minute addition to your crew's end-of-job checklist. Maybe they're running Google Ads and you're not — a conversation to have with your marketing setup.

Small adjustments compound. The business that runs a competitive review twice a year and acts on three things each time will outpace the one that never looks up from its own operation.

Build it into your calendar. The same way you review your P&L monthly, you review your competitive landscape quarterly. Make it a habit and it stops feeling like extra work — it just becomes part of how you run the business.

> *Stay informed. Stay sharp. The operators who know their market win it.*

Go Deeper

For more on this and everything else in the book, the Haulers' Edge Newsletter goes deeper every week. Scan the code or subscribe free at HaulingHubb.com.

CH. 8 STOP CHASING EVERY JOB

> *"There is only one winning strategy: carefully define a target market and direct a superior offering to that target market." — Philip Kotler*

Every service business owner figures this out eventually — usually the hard way. You say yes to everyone, run yourself into the ground trying to be everything to everybody, and end up with a schedule full of jobs that barely cover costs and customers who drain you.

The fix isn't working harder. It's getting honest about who you actually want to serve — and building everything around them.

This isn't a marketing theory exercise. Knowing your ideal customer changes how you price, how you route, how you hire, what you say in ads, and how your crew talks to people on-site. It touches every part of the operation. Get it right and the business gets easier. Stay vague about it and you'll keep grinding for mediocre results no matter how many hours you put in.

Who You're Actually Talking To

Most service businesses have two or three distinct customer types, each with completely different needs, different budgets, and different reasons for calling.

In hauling and junk removal the clearest split is residential versus commercial. A homeowner cleaning out a basement wants reliability, a friendly crew, and peace of mind that the job gets done without drama. A contractor managing a job site wants fast dumpster swaps, no delays, and someone who picks up the phone at 7am. Those are different people. They respond to different messages. They need different things from you.

A landscaper has the same split — the homeowner who wants a beautiful yard versus the property management company that needs 40 units maintained on a schedule. A cleaning company serves the residential client who wants their house to feel like home and the commercial client who needs an office turned over before 8am. Same trade, completely different customer.

The mistake most operators make is trying to market to both with the same message. It doesn't work. Generic messaging that tries to appeal to everyone connects with no one. Specialists inspire trust. Generalists get compared on price.

Pick the customer types you want most. Build your pitch around them. You'll still get other work — but you'll be known for something specific and that reputation compounds.

Build a Real Picture of Each Customer Type

A customer profile — sometimes called a persona or avatar — is just a detailed description of a typical customer you want to attract. Not a fictional character for its own sake. A tool that makes every marketing and operational decision easier because you're making it for a specific person instead of an abstraction.

For each customer type you want to target, work through these dimensions.

Demographics. Who are they practically? Age range, location, occupation, general income level. A 45-year-old homeowner in the suburbs and a 38-year-old general contractor have completely different lives and completely different relationships with your business. If you serve both residential and commercial customers, build a separate profile for each — they'll diverge significantly across every dimension.

What they actually care about. This is the one that changes your marketing. Some customers care most about convenience — they want it done fast with minimal involvement from them. Some care most about reliability — they've been burned before and want to know you'll show up when you say you will. Some care about eco-friendly disposal. Some care about price. You can't know which matters most unless you think it through — and once you do, your messaging becomes specific instead of generic.

Their pain points. What problem are they actually trying to solve? The homeowner isn't calling because she wants junk removed. She's calling because the garage has been unusable for two years and it's been stressing her out every time she looks at it. The contractor isn't calling because he needs a dumpster. He's calling because his last hauler was unreliable and it cost him a week on a flip. Understand the real problem underneath the service request and you can speak to it directly.

What triggers the decision to call. Homeowners often act around life transitions — moving, a death in the family, a renovation, spring cleaning. Contractors call when a project starts or when a current hauler lets them down. Property managers call when a tenant turns over. Knowing the trigger tells you when to be in front of them and what to say when you are.

How they prefer to be reached. Homeowners often find you through Google search, Nextdoor, or a neighbor's recommendation. Contractors respond to direct outreach, trade associations, and word of mouth from other contractors. Younger customers expect to book online and get a text confirmation. Older customers want to talk to a person. If you're showing up in the wrong place or communicating in the wrong way, the right customers never find you even if you're the best option in the market.

Three Customer Types Worth Knowing

Here are three profiles that show up consistently across service businesses. These are composites — not every customer fits perfectly — but they're realistic enough to be useful.

Carol the Homeowner. Mid-40s, suburban homeowner, years of accumulated clutter she's been meaning to deal with. She's slightly embarrassed to ask for help and wants to feel like she made a smart choice. She's not necessarily price-driven but she wants to feel like she got fair value. She finds you on Google or through a neighbor's recommendation. What closes her: a friendly, professional interaction on the phone, clear pricing with no surprises, and reassurance that your crew will treat her home with respect. What loses her: an aggressive sales pitch, a vague price range, or a crew that shows up looking unprepared.

Gary the Contractor. Late 30s, running job sites and flipping properties. His currency is time. Every day a project sits because a dumpster didn't show up or debris is blocking the work area is money out of his pocket. He doesn't care about eco-friendly disposal or friendly crews. He cares that you pick up the phone, deliver when you say you will, and solve his problem without creating a new one. He finds you through word of mouth from other contractors or direct outreach. What closes him: demonstrating that you're reliable and easy to work with. What loses him: showing up late once.

Ashley the Urban Professional. Late 20s to early 30s, renting or in a condo, tech-comfortable and environmentally conscious. She has a small project — a cleanout, a dumpster bag for a renovation — and she wants to handle it entirely from her phone without a single phone call if possible. She'll read your reviews before she books. She cares whether items get donated or recycled. What closes her: clean mobile-friendly website, online booking, transparent pricing, and visible social proof. What loses her: a clunky booking process or a phone-only operation.

The practical implication: your Google Ad targeting, your website copy, your phone script, and your on-site crew behavior should all reflect which of these customers you're talking to in that moment. One conversation, one message, one customer type at a time.

Use Your Existing Data Before You Guess

Before you build profiles from scratch, look at what your business already knows.

Pull your last 50 jobs and categorize them. What types of customers appear most often? Which ones generate the highest revenue per job? Which ones have the best margins when you account for True Job Cost? Which ones refer other customers? Which ones create the most headaches?

That analysis will almost always show you that a small number of customer types are driving a disproportionate share of your revenue and profit. According to Bain & Company research, increasing customer retention by just 5% can increase profits by 25–95% — figures drawn from financial services, but the directional point holds for any business where repeat customers cost less to serve and refer more freely *(Bain & Company, 2000)*. You're not looking for more customers. You're looking for more of the right customers.

Talk to the ones you already have. Ask your best customers directly — why did they call you, what made them book, would they refer you and why. The answers are more useful than anything you'll theorize from a spreadsheet. In my own business, customer feedback revealed demand for a service line I hadn't been marketing — it later became a meaningful profit center.

Aim Your Marketing at the Right People

Once you know who you're targeting, your marketing gets dramatically more efficient because you stop paying to reach people who were never going to buy.

On paid ads, use what you know about each persona to tighten your targeting. Google search ads should use the language your customer type actually types — a homeowner searches "junk removal near me," a contractor searches "dumpster rental for contractors" or "same-day roll-off delivery." Different terms, different intent, different ad. Running one generic ad for both audiences means your message is slightly wrong for everyone.

On your website, create distinct sections or pages for your primary customer types. Residential customers want to see friendly crews, clear pricing, and reassurance. Commercial customers want to see capacity, reliability, and fast turnaround. The same company, two different conversations. When someone lands on the page that speaks their language, your conversion rate climbs.

On social media, let your personas guide your content. A before-and-after garage cleanout photo speaks to Carol. A dumpster delivered and picked up on the same day with zero job site delay speaks to Gary. Don't post for everyone — post for the specific person most likely to need you this week.

Your email and text outreach should be segmented the same way. Homeowners get a different message than contractors. Property managers get a different message than both. A segmented email to the right audience consistently outperforms a generic blast to everyone — and it takes the same amount of time to write once you have the lists organized.

A CRM — even a simple one — makes this segmentation practical rather than aspirational. Tag customers by type when they book. Filter by tag when you send a campaign. Track which customer types respond to which messages. Over time you build a picture of what works for each segment and you stop guessing. We use Service Hubb AI at Grizzly Junk Pros and it handles all of this in one place — booking, follow-up, segmentation, and automated text campaigns by customer type. It's built specifically for service businesses, so the pipeline and automation are already structured around how a service operation runs.

Test, Measure, Adjust

Your first set of customer profiles is a hypothesis. The data tells you whether you're right.

Run two versions of an ad with different messages — one emphasizing speed, one emphasizing reliability — and see which gets more clicks and conversions from the audience you care about. If your Google Ads are filling your schedule with small residential pickups but you're targeting contractors, your keywords or targeting are off. If your email open rates are high from one customer type and low from another, the content is wrong for one of them.

Ask every new customer how they found you and what made them call. Track the answers. Patterns emerge quickly. If three customers in a row found you through a neighbor's recommendation after seeing your truck, that's a signal to put your brand on everything — trucks, uniforms, yard signs. If every contractor says the same competitor let them down and they were looking for an alternative, that's a story to tell in your marketing.

Update your profiles as you learn. Maybe you assumed price was the primary concern for residential customers and testing reveals reliability matters more — they'll pay a premium for someone who shows up when they say they will. That single insight changes your entire ad message and how your crew answers the phone.

The goal is to move from guessing to knowing. Not immediately — it takes a few cycles of testing and adjustment. But every iteration gets you closer to a calendar full of the right jobs: profitable, repeatable, and the kind of work your operation is actually built to do well.

If you already know who your best customer is and you've been saying no to the wrong jobs for years, the question isn't who — it's whether your marketing, pricing, and intake process are aligned tightly enough that the wrong customers stop calling entirely. The goal isn't to filter better. It's to build a system where the wrong leads never reach you in the first place: ad targeting specific enough that off-profile customers don't click, minimum job requirements that weed out small-ticket requests before they reach the phone, and a reputation in the right channels that attracts the customer type you want without you spending money reaching the ones you don't. Knowing your customer is the first step. Engineering your entire operation around them is the next one.

> *Stop chasing every job. Start attracting the right ones. The difference shows up in your margins, your schedule, and how you feel at the end of the week.*

Go Deeper

For more on this and everything else in the book, the Haulers' Edge Newsletter goes deeper every week. Scan the code or subscribe free at HaulingHubb.com.

CH. 9 BECOME THE NAME THEY THINK OF FIRST

"Your brand is what people say about you when you're not in the room." —Jeff Bezos

There's a moment that separates the businesses that grind for every job from the ones that have work coming to them consistently. It's the moment when someone in your market — a homeowner, a contractor, a property manager — needs what you do and your name is the first one that comes to mind. Not because you ran an ad that week. Not because you were the cheapest option on Google. Because your name is just... there. Familiar. Trusted. The obvious call.

That's what a local brand does. And most service business owners have no idea they're building one — or failing to — with every interaction, every truck on the road, every review, and every job they do or don't show up for on time. The operators who win long-term are the ones who became the most known and most trusted name in their market.

This chapter is about building that recognition deliberately so it compounds over time instead of leaving it to chance.

Brand Is Not Your Logo

Most people hear "brand" and think logo, colors, maybe a tagline. Those things matter but they're the surface. Your brand is the full impression your business leaves on everyone who encounters it — before, during, and after the job.

It's whether your truck looks professional or like it rolled in from a salvage yard. It's whether your crew shows up in clean shirts or whatever they grabbed that morning. It's whether your phone gets answered on the second ring or goes to a voicemail that's never been set up. It's your Google review average and how you respond when someone leaves a bad one. It's whether the job site looks better when you leave than when you arrived. It's whether the customer hears from you after the job or you disappear until they need you again.

Every one of those touchpoints is a brand impression. Managed well, they stack into a reputation that makes your name valuable in your market. Ignored, they leave customers with no particular reason to choose you over anyone else — or to remember you at all.

The good news for small service businesses: you don't need a big budget to build a strong local brand. You need consistency. Consistency at scale, repeated over time, is what makes a name stick.

If you're already established and think brand-building is what you did in years one through three, the work has changed — not stopped. Early brand-building is about visibility: getting your name in front of the right people often enough that it sticks. At $400K+ in revenue, the shift is toward authority: being known not just as a business that shows up but as the operator others in your market reference, learn from, and compare themselves to. That looks like being the company that publishes pricing when nobody else does, the one contractors mention by name to other contractors, the one with a Google review profile so dominant that new entrants have to acknowledge it. Visibility gets you found. Authority gets you chosen without comparison. The tactics are different. The compounding still applies.

Define What You Want to Be Known For

Before you can build recognition, you need to be clear on what you want people to recognize you for. Not a vague list of virtues — every business claims to be reliable, professional, and customer-focused. Something specific. Something that reflects how you actually operate and what genuinely differentiates you.

Ask yourself honestly: when a customer who used you tells a neighbor about their experience, what do they say? Not what you wish they said — what they actually say. That's your current brand reality. It might be exactly what you want. It might show a gap between how you see yourself and how the market sees you.

For Grizzly Junk Pros, the answer is fast, reliable, no-pomp and circumstance service. We're not the cheapest option and we don't try to be. We show up when we say we will, the crew is professional, and the job gets done right the first time. That's what we want people to say. Everything about how we operate — from how we answer the phone to what the trucks look like to how we follow up after a job — is designed to reinforce that.

For your business, pick one or two things. Speed. Reliability. Eco-friendly disposal. Deep local roots. Contractor specialization. Whatever is both true and differentiated. Then build everything outward from that core.

Three questions that point you to the right answer: What do your five-star Google reviews specifically mention — not the rating, the actual words customers use? What do repeat customers say when you ask them why they came back? And what do you do consistently that the top three competitors in your market don't advertise or offer? The overlap between those three answers is your brand position. It's already there in your operation — you just haven't named it yet.

If every competitor in your market is positioning on price and you can genuinely deliver faster and more reliably, speed is your brand. If everyone is generic and you've been in the community for a decade, local trust is your brand. If you're the only operator in your area who donates usable items and recycles aggressively, sustainability is your brand. The goal is to own a position that matters to your target customer and that your competitors either can't or haven't claimed.

The business willing to be more transparent, more human, and more open than every competitor in its market doesn't just win visibility — it wins trust that no amount of advertising can replicate. At Grizzly Junk Pros, transparency isn't a marketing strategy — it's just how we operate. Customers know what the dump costs. They know what our trucks and containers cost to run. They know how expensive commercial insurance coverage is. They know our team is paid well above minimum wage. That openness doesn't lose us jobs — it wins them. The customer who understands what they're paying for and why is a different conversation than the one wondering if they're being taken advantage of.

Make Your Name Visible Everywhere It Can Be

Brand recognition is built through repetition. People need to see or hear your name multiple times before it sticks — research in marketing consistently points to the concept of seven or more touchpoints before a prospect converts *(Marketing Rule of 7, widely cited)*. In a local service market, that repetition happens through physical presence as much as digital.

Your trucks are your most powerful brand asset and most businesses underuse them. A clean, well-branded truck driving through your service area every day is a moving billboard. It shows up at job sites, in driveways, at gas stations, in traffic. Every neighbor who sees it on a job site has a visual reminder that someone in their area uses you. That passive exposure compounds over months and years into name recognition that no ad can replicate.

The truck needs to be clean and the branding needs to be legible. Your business name, your phone number, your website. Large enough to read from a moving car. That's it. It doesn't need to be elaborate — it needs to be consistent and visible.

Yard signs at job sites work the same way. A sign in the front yard during a cleanout or renovation job puts your name in front of every neighbor walking or driving by. Get permission, place the sign, pick it up when the job is done. A small cost with disproportionate local visibility return. If customers hesitate to give up the space, a small incentive removes the friction — offer $20 off their service in exchange for leaving the sign up until they mow the lawn. Most people say yes, the sign stays up longer than the job takes, and you've just bought a week or two of neighborhood visibility for twenty dollars.

Uniforms matter more than most owners think. A crew in matching shirts with your company name reads as professional. It signals to the customer that you run a real operation, not a truck and a guy. It also protects your brand on job sites — when your crew looks the part, people assume the work will match.

Your Google Business Profile Is Your Storefront

For most local service businesses, more potential customers will form their first impression of you through your Google Business Profile than any other single touchpoint. It shows up when someone searches your service in your area. It shows your reviews, your photos, your hours, and how you respond to feedback. It's often the deciding factor in whether someone calls you or the next result.

Most operators set it up once and forget it. That's leaving serious brand value on the table.

Keep it current. Your hours, your service area, your phone

number, your website — all accurate and up to date. Add photos regularly. Real photos of your crew, your trucks, completed jobs. Not stock images. Real images build trust in a way stock photos never will because they show the customer exactly what to expect. A profile with twenty real job photos converts better than one with two stock images of a smiling handyman.

Your business description should be specific and use the language your customers actually search. Not "we provide comprehensive waste management solutions" — nobody searches that. "Junk removal and dumpster rentals serving homeowners and contractors in [your city and surrounding towns]." Specific, searchable, and immediately clear about who you serve.

Post updates regularly — Google rewards active profiles with better visibility. A before-and-after photo from a job this week. A seasonal reminder that spring cleanout season is coming. A note about a new service. Ten minutes once a week keeps the profile active and signals to Google's algorithm that your business is current and engaged.

Reviews Are Your Brand Currency

In a local service market, your review profile is your reputation made visible. Your star rating and review volume determine whether a prospect who finds you on Google feels confident calling or keeps scrolling.

The number one mistake service businesses make with reviews is passive. They do good work and hope customers leave reviews on their own. Some do. Most don't — not because they're unhappy, but because it didn't occur to them and the moment passed. The businesses with 200 reviews aren't getting them accidentally. They're asking for them systematically.

Ask every satisfied customer for a review before the crew

leaves the job site. Not a generic "if you could leave us a review that would be great" — a specific, direct ask. "We'd really appreciate it if you left us a review on Google — it takes about a minute and it helps us a lot. Do you want me to text you the link?" Most people say yes. Most of those people follow through if you make it frictionless.

Automate the follow-up. A text or email with a direct link to your Google review page, sent within an hour of job completion, catches customers while the experience is fresh. At Grizzly Junk Pros this is built into our workflow — the job closes, the follow-up goes out automatically. The reviews compound.

Respond to every review — positive and negative. Responding to positive reviews shows you're engaged and grateful. More importantly, it shows potential customers reading those reviews that a real person is behind the business. Responding to negative reviews professionally — acknowledging the concern, offering to make it right, not getting defensive — shows how you handle problems. A business with a 4.7 average and professional responses to its few negative reviews often converts better than a business with a 4.9 average and no responses at all, because the responses show character.

Never respond defensively or argumentatively to a negative review in public. No matter how wrong the customer is, the argument plays out in front of everyone who reads it. Take the high road, offer to resolve it offline, and move on.

Consistency Builds Trust Over Time

The businesses that become the name people think of first in their market don't usually have one brilliant marketing move. They have a hundred consistent ones, repeated over months and years until the name is embedded in the community.

The truck shows up clean every day. The crew wears the shirts. The phone gets answered. The follow-up text goes out. The review link gets sent. The Google profile gets updated. The yard sign goes in at every job. None of these is a home run on its own. Together, over time, they create a brand that feels familiar and trustworthy — and familiar trust is what makes someone call you without shopping around.

Inconsistency is the enemy of brand building. One crew member who doesn't wear the uniform. One week where follow-up texts don't go out. One month where the Google profile sits stale. These gaps don't kill a brand immediately, but they slow the compounding. Consistency is a habit, not a one-time decision.

Build the habits into systems so they don't depend on you remembering. Crew uniform policy in the employee handbook. Follow-up text automated in the CRM. Google profile update scheduled on the calendar. The system runs the consistency so you don't have to.

Your Community Presence Is Part of Your Brand

For local service businesses, being visible in the community you serve has a brand value that digital marketing can't replicate. People want to hire companies that feel like their neighbors, not faceless operations that happen to service their zip code.

Sponsor the local little league team. Your name on the shirts gets seen at every game by every parent — a highly local, highly qualified audience. The cost is minimal and the community goodwill is real. Join the Chamber of Commerce not just for the networking but for the visibility in local business directories and events. Show up at community clean-up days. Partner with local charities on donation pickups — it generates goodwill, gives you something genuine to talk about in your marketing, and differentiates you from competitors who just haul everything to the dump.

None of this needs to be elaborate or expensive. It needs to be consistent and genuine. A company that shows up in the community year after year builds a kind of trust that no Google Ad can buy.

When you do good work in the community, talk about it. Post the before-and-after from the charity cleanout. Tag the organization. Share the story. Community involvement only builds your brand if people know about it — sharing it isn't bragging, it's evidence of who you are. The businesses that feel the most human in a market full of faceless operations win a loyalty that visibility alone can never buy — show your face, introduce your crew, and let people see who they're actually hiring before they call. Video is the fastest way to show that evidence — a two-minute phone video of a real job or a community event does more for brand recognition than a month of static posts, and most of your competitors aren't doing it. Full breakdown in Chapter 13.

The Long Game

Brand building is not a campaign. It doesn't have a start and end date. It's the aggregate of every interaction, every impression, and every touchpoint over the life of your business. The operators who invest in it consistently over years end up with something competitors genuinely cannot buy — a name that means something in their market.

When your brand is strong enough, marketing gets easier. Customers find you before they find competitors. They come in pre-sold because someone they trust recommended you. They book without getting three quotes. They leave reviews because they feel connected to the business, not just transactionally satisfied.

That's not luck. That's what consistent brand building produces over time.

Start with the basics. Clean trucks, professional crew, fast follow-up, systematic reviews. Get those right before you worry about anything else. Then layer in community presence, content, and visibility as the capacity to do so grows.

> *The name people think of first in your market didn't get there by accident. Build it deliberately or watch someone else take the position.*

The visibility metrics that tell you whether your brand-building is working — GBP views, website sessions, review velocity — are in your weekly scorecard in Appendix A.

Go Deeper

For more on this and everything else in the book, the Haulers' Edge Newsletter goes deeper every week. Scan the code or subscribe free at HaulingHubb.com.

CH. 10 ONE BAD REVIEW CAN BURY YOU

"It takes 20 years to build a reputation and five minutes to ruin it." — Warren Buffett

You've been building your business for years. Good work, happy customers, a name people in your market are starting to recognize. Then one morning you open your phone and there it is. One star. A paragraph that makes your stomach drop. Maybe it's exaggerated. Maybe it's completely unfair. Maybe the customer was a problem from the first call and you knew this was coming.

Doesn't matter. It's public. It's permanent until you address it. And every potential customer who finds you on Google is going to read it before they decide whether to call.

This chapter is about understanding how much reviews actually affect your business, how to respond when things go wrong, how to protect your reputation proactively, and how to build a review profile strong enough that one bad one doesn't define you.

How Much Reviews Actually Matter

The data is unambiguous. According to BrightLocal's 2023 Local Consumer Review Survey, 98% of consumers read online reviews for local businesses and 87% say reviews are the primary factor in their decision to contact a business *(BrightLocal, 2023)*. For a local service business competing on Google, your review profile is not a nice-to-have. It's the front door.

A single one-star review doesn't just affect how customers feel about you — it affects whether they can find you at all. Google's local search algorithm factors in review quantity, recency, and average rating when determining which businesses to show in the local pack — the three businesses that appear at the top of a local search result. A sudden drop in rating or a cluster of negative reviews can push you down the results page and hand those positions to competitors. Lower visibility means fewer calls. Fewer calls means less revenue. The downstream effect of a bad review handled poorly is larger than most owners expect.

According to a Harvard Business School study, a one-star increase in Yelp rating leads to a 5–9% increase in revenue for the average business *(Harvard Business School, 2011)*. The inverse is equally true. One bad review, left unaddressed, sitting at the top of your profile, is costing you jobs every single week.

Why Bad Reviews Happen to Good Businesses

The first thing to understand is that negative reviews are inevitable. If you're doing enough volume and serving enough customers, you will eventually get one. Not because you failed — because you're in a business that involves people, logistics, and expectations that don't always align.

The most common sources of negative reviews in service businesses are not catastrophic failures. They're gaps between expectation and reality. The customer expected a two-hour job and it took four. The price came in higher than the phone estimate because the job was bigger than described. The crew was professional but quieter than the customer expected. The dumpster got placed two feet from where the customer visualized it.

None of those are inexcusable. All of them can trigger a one-star review if the customer feels unheard or if nobody caught the issue before they got home and opened Google.

The second source is customers who were problems from

the start — unrealistic expectations, price-shoppers who were never satisfied with what they paid, people who had a bad day and took it out on the review. These reviews sting more because they feel unfair. The response strategy is the same regardless.

The third source — and the most preventable — is a genuine service failure. The crew showed up late with no communication. Something got damaged and nobody acknowledged it. A job was quoted one price and billed another. These are operational failures and the review is a symptom of a system problem, not just a PR problem. Fix the system.

How to Respond to a Negative Review — The Right Way

Your response to a negative review is not primarily for the person who left it. It's for every potential customer who reads it afterward. How you handle public criticism tells people more about your business than the complaint itself.

The right response has four parts. Acknowledge, empathize, offer to resolve, take it offline.

Acknowledge what they experienced without immediately defending yourself. Not an admission of fault — an acknowledgment that their experience didn't meet expectations. "I'm sorry to hear your experience didn't go as expected" is honest and disarming. "We always deliver excellent service and I'm not sure what happened here" is defensive and makes you sound like you're already arguing with the customer.

Empathize briefly. One sentence that shows you understand why they're frustrated. "I completely understand how frustrating an unexpected delay can be, especially when you've cleared your schedule for the job."

Offer to resolve it. Make clear you want to make it right and give them a direct way to reach you. Not a general "contact us" — your name and a direct number or email. "I'd like to personally understand what happened and make this right. Please call me directly at [number] or email [address]."

Keep it short. Three to five sentences maximum. A long response reads as defensive even when the content is reasonable. Say what needs to be said and stop.

What you never do: argue with the review, point out that the customer was wrong, list all the reasons the complaint is unfair, or respond sarcastically. All of that plays out in front of everyone who reads it and makes you look worse than the review itself. Even when you're completely right, the public argument is a loss.

Here's an example of what a right response looks like in practice.

A customer leaves a one-star review saying your crew showed up an hour late and didn't call ahead.

Wrong response: *"We reached out multiple times and the crew was only 45 minutes late due to traffic on a job that ran long. We communicated this as best we could and completed the job professionally."*

Right response: *"I'm sorry we didn't meet your expectations on timing and communication — you deserved better than that. Punctuality matters to us and clearly we fell short here. I'd like to personally make this right. Please reach out to me directly at [number]. —Justin"*

The right response acknowledges the specific complaint, takes ownership without over-explaining, offers resolution, and ends with your name. Personal. Accountable. Brief.

Here's the counterintuitive truth about negative reviews that most operators miss. A business with 200 five-star reviews and zero negative responses looks curated — like someone is managing the image rather than running a real operation. A business with 195 five-star reviews, three negative reviews, and three genuine human responses that take accountability and offer resolution looks real. Customers reading those responses aren't thinking "this company has problems." They're thinking "this is a company that handles problems like an pro." That distinction converts skeptical browsers into callers faster than another five-star review does. The negative review you respond to honestly is doing sales work for you in front of every potential customer who reads it. Treat it accordingly.

When the Review Is Genuinely False or Malicious

It happens. A competitor leaves a fake review. A former employee with a grudge posts something fabricated. Someone who never used your business somehow finds their way to your profile.

Your first move is to report it to Google using the flag function on the review. Google will investigate and remove reviews that violate their policies — reviews from people who were never customers, reviews that contain hate speech, competitor-planted reviews. The process takes time and Google doesn't remove every flagged review, but it's always worth attempting for reviews that are clearly fraudulent.

While you're waiting on that process, respond publicly the same way you would a legitimate complaint — calmly, briefly, professionally. Something like: "We don't have any record of a job matching this description in our system. We take all feedback seriously and would appreciate the opportunity to understand what happened. Please contact me directly at [number] so we can look into this." This signals to readers that the review may not be legitimate without getting into an accusatory public argument.

Document everything. If you have records proving the reviewer was never a customer — no booking, no job, no transaction — save that documentation. If the fake review situation escalates, having records supports your case with Google and, in rare situations, legal action.

What you should never do with a suspected fake review is respond accusatorially in public. "This is clearly a fake review from a competitor" might be true but it reads as paranoid defensiveness to anyone who doesn't have the full context.

The Review You Can Prevent

The best review management strategy is catching problems before they turn into public complaints.

Most customers who have a bad experience don't immediately open Google. They stew about it. They tell a neighbor. They decide they'll never use you again but they don't take any action until something triggers it — usually seeing your truck again, or getting a survey email, or having a second bad experience with someone else that reminds them of yours.

That window — between when the experience happened and when the review gets posted — is your opportunity.

A follow-up call or text after every job does two things. It catches unhappy customers before they go public, and it gives satisfied customers a natural moment to leave a review. Make it personal, not a form letter. "Hey, this is Justin from Grizzly Junk Pros — just wanted to make sure everything looked good after yesterday's cleanout. Let me know if there's anything we missed." That one message resolves more problems than any review response strategy.

Train your crew to read the job. If they sense a customer is unsatisfied at the end — quieter than expected, not engaging, looks like they're biting their tongue — they should flag it before they leave. Not ignore it and hope for the best. A simple "Is there anything we can do to make this better before we head out?" catches most issues in the moment when they're still fixable.

Build a feedback mechanism into your booking process. A short survey after job completion — two or three questions, one to five stars, an optional comment field — gives dissatisfied customers a private channel to voice concerns. If someone rates you two stars on the internal survey, you call them before the public review goes up. That interception rate is high when the process runs smoothly.

Build a Review Profile That Absorbs Bad Ones

The most durable protection against a bad review is a review profile strong enough that one bad one doesn't define you. A business with 8 reviews is devastated by a one-star. A business with 200 reviews absorbs it.

The mechanics of generating reviews consistently — asking every satisfied customer, automating the follow-up text with a direct link, making it frictionless — are the foundation. The additional principle that separates good from great is recency. Google weights recent reviews more heavily than old ones in its local ranking algorithm. A business with 200 reviews that's been dormant for a year is more vulnerable than one with 80 reviews that are consistently recent.

The goal is a steady flow of reviews over time, not a burst campaign. A burst campaign — where you ask every past customer at once — generates a spike of reviews that looks unnatural and can actually trigger Google's spam filters. Consistent, ongoing review generation from current customers is both more natural and more algorithmically durable.

Respond to positive reviews too. Not just "thanks!" — a brief, genuine response that references something specific about their experience. "So glad the garage cleanout went smoothly — that was a big one. Thanks for trusting us with it." These responses show prospective customers that you're engaged, personal, and that real interactions happen between you and the people you serve.

A strong review profile with consistent recent reviews and active owner responses is one of the most valuable assets your local brand has. It takes time to build and is genuinely difficult for a competitor to replicate quickly.

The Two-Path Review System

Asking every customer for a review and sending them directly to Google is the most common approach. It's also the most naive. Not every customer who had a good experience will leave a five-star review. And not every customer who had a frustrating moment will tell you before they tell the internet. The Two-Path Review System fixes both problems.

I know what you might be thinking — didn't I just say negative reviews have value? They do. This isn't about burying legitimate feedback or pretending problems don't exist. It's about sequence. A customer who feels unheard rushes to Google. A customer who gets a direct line to you and feels genuinely heard often doesn't — and when they do post a review later, the experience of being handled well changes the tone. You're not suppressing the feedback. You're giving it a better first destination.

Here's how it works at Grizzly Junk Pros. After every completed job, an automated follow-up goes out — a text and an email — asking the customer to share their experience. But instead of sending them directly to Google, the first step is a single question: are you satisfied or unsatisfied with your service today?

If they click satisfied, they're routed to your review destination — typically Google, which carries the most weight for local search ranking. But not every satisfied customer has a Gmail account, and without one they can't leave a Google review. So the satisfied path includes a secondary option — a landing page on your website that lists every high-authority platform where a review helps your business. Google first, but also Yelp, Facebook, the Better Business Bureau, and any industry-specific directory that matters in your market. The customer who can't leave a Google review can still leave one somewhere that counts, and you capture the positive sentiment rather than losing it.

If they click unsatisfied, they never see a public review platform. Instead they're routed to an internal feedback page — a private form where they can describe what went wrong, express their frustration, and feel genuinely heard. That feedback stays inside your system. It doesn't get posted anywhere. The customer got to air their grievance. You get to see exactly what happened and reach out to make it right before the situation escalates.

This matters because a customer who feels unheard becomes a one-star review. A customer who feels heard — even if the experience wasn't perfect — often becomes a loyal one. The Two-Path Review System doesn't suppress legitimate feedback. It creates a channel for that feedback to reach you directly instead of reaching Google first.

The business benefits stack up quickly. Your public review profile only receives input from customers who self-identified as satisfied. Your internal system captures every complaint before it goes public and gives you a direct line to resolve it. And customers who might not have left any review — because they weren't angry enough to complain publicly but weren't thrilled enough to volunteer a five-star — get a frictionless path to leaving positive feedback on the platform of their choice.

Service Hubb AI handles this entire sequence automatically — the post-job trigger, the satisfaction question, the routing logic, the internal feedback capture, and the follow-up if a complaint comes in. Once it's configured it runs on every job without manual involvement. The system works while you're on the next job.

Build this before you need it. Because the review that would have cost you three future customers never gets posted — and the customer who almost left angry becomes the one who tells their neighbor you handled it right.

When a Review Reveals a Real Problem

Sometimes a negative review isn't just a PR problem. It's a signal.

If you get three reviews in two months that all mention the same crew member being rude — that's a management issue, not a review issue. If multiple reviews mention pricing surprises at the end of a job — that's a quoting and communication process issue. If recurring complaints cluster around scheduling or punctuality — that's an operations issue.

Treat clusters of similar negative feedback the way you'd treat a pattern in your P&L. It's data telling you something is broken. The review is a symptom. Fix the system that caused it and the reviews stop coming.

This is actually one of the most valuable functions of your review profile. It's real-time feedback on your operation from people who have no incentive to be polite about it. Most owners treat negative reviews as a reputation problem to manage. The ones who build the best businesses treat them as an operational diagnostic tool.

When you make a change in response to review feedback — adjusted your quoting process, retrained a crew member, changed how you communicate delays — and the complaints stop, you've confirmed the fix worked. That feedback loop, used honestly, makes your business better faster than most internal quality control processes.

Platform Priorities

Google is the primary platform for most local service businesses and should receive the most attention. It directly impacts your search visibility and is where the majority of local purchase decisions get made.

Facebook reviews matter for the segment of your customer base that finds you through social media. Keep your page active and respond to reviews there with the same discipline as Google.

Yelp has a complicated relationship with small businesses — their algorithm sometimes filters legitimate positive reviews and their ad sales tactics are aggressive. Maintain a basic presence, respond to reviews, but don't build your strategy around it.

Angi and HomeAdvisor matter if you generate leads through those platforms. Reviews there affect your standing in their marketplace and are worth managing.

Nextdoor is increasingly important for residential service businesses — it's where neighbors ask for recommendations and where your name can spread organically through word of mouth at scale. Encourage happy customers who are active on Nextdoor to mention you when neighbors ask for service recommendations. That organic word-of-mouth is the highest-trust recommendation in the local market.

Focus your energy on Google first. Get that right, then layer in the others.

> *Your reputation is being built in public every day whether you're managing it or not. The question is whether you're the one shaping it.*

Review velocity — how many new reviews you're generating each week — is one of the weekly metrics in Appendix A.

Go Deeper

For more on this and everything else in the book, the Haulers' Edge Newsletter goes deeper every week. Scan the code or subscribe free at HaulingHubb.com.

CH. 11 YOUR BEST LEAD IS ALREADY A CUSTOMER

> *"There is only one boss: the customer. And he can fire everybody in the company from the chairman on down, simply by spending his money somewhere else." — Sam Walton*

The most expensive thing your business does every day is find new customers. Ads, SEO, follow-up sequences, time on the phone, free estimates — all of it costs money and energy to convert a stranger into someone who trusts you enough to book.

You've already done that work with every customer who's ever used you. They know your name. They've seen your crew. They have evidence that you show up and do the job right. Selling to them again costs a fraction of what it cost to acquire them in the first place — and yet most service business owners spend almost all their marketing energy chasing strangers while their existing customer base quietly goes cold.

That's backwards. Your best lead is already in your CRM.

The Numbers Make the Case

Acquiring a new customer costs five to twenty-five times more than retaining an existing one (Harvard Business Review, 2014). That single number should change how you think about your marketing budget. Your probability of closing a past customer runs 60–70%, compared to 5–20% for a new prospect (Marketing Metrics, Paul Farris et al.). And a 5% improvement in customer retention can increase profits by 25–95% (Bain & Company, 2000).

Run that math on your own business. If you generate $100,000 in monthly revenue at a 10% net margin, that's $10,000 in profit. A 5% retention improvement that produces even a 25% profit increase adds $2,500 a month — $30,000 a year — without a single new customer, without running a new ad, without hiring anyone. Just from keeping more of the customers you already have and getting them to come back.

The customers you've already served are the highest-ROI marketing channel your business has. Most operators treat them like the end of a transaction. The ones who build real businesses treat every completed job as the beginning of a relationship.

Why Customers Don't Come Back — And It's Usually Not What You Think

Most lost repeat customers didn't have a bad experience. They had a fine experience. They just forgot about you.

Life moves fast. A homeowner you did a garage cleanout for in April isn't thinking about you in October when leaves pile up in the yard and the basement starts filling again. Not because she was unhappy. Because you're not in front of her. Another company's truck drove by. A neighbor mentioned a different name. She Googled it and found someone with more recent reviews.

You lost a repeat customer not to a competitor but to irrelevance. And irrelevance is entirely preventable.

The second reason customers don't come back is they don't know you do everything you do. The homeowner who hired you for a junk cleanout doesn't know you also rent dumpsters. The contractor who rented a dumpster doesn't know you do demolition. You solved one problem for them and they filed you under that category — forever. They'll call a different company for the next problem because it didn't occur to them that you could handle it.

Retention starts with staying visible and staying relevant. That's it. Not complicated. Just consistent.

Build the System That Does It Automatically

The operators who retain customers well almost never do it through sheer effort and memory. They build a system and the system runs it. Here's what that system looks like in practice.

Immediate follow-up — within 24 hours of every completed job. A text or email that thanks them, confirms they're satisfied, and includes a direct link to leave a Google review. Keep it short and personal, not a corporate form letter. "Hey Mrs. Lopez — wanted to make sure everything looked good after yesterday's cleanout. If you have a minute to leave us a review it really helps: [link]. And we're here anytime you need us. — Justin." That message does three things in thirty seconds: confirms satisfaction, generates a review, and plants the seed of the next job.

Thirty-day follow-up. A short check-in or a helpful piece of content. "Spring cleaning season is coming up — let us know if the garage is starting to fill again." Not a pitch. A reminder that you exist and you're thinking about them.

Seasonal touchpoints. Service businesses have natural seasonal rhythms. Lean into them. A late winter message about spring cleanouts. A fall message about getting ahead of year-end projects. A post-holiday message about clearing out what didn't fit. These aren't aggressive sales emails — they're timely reminders from someone they already trust that show up at exactly the moment the need is starting to form.

Annual outreach to dormant customers. Anyone who used you more than twelve months ago and hasn't been back gets a personal message. “It's been about a year since we helped you clear out the basement — wanted to check in and see if you have anything piling up again. Here's 10% off if you book this month.” Some of these won't respond. A meaningful percentage will. All of it costs almost nothing.

At Grizzly Junk Pros this whole sequence runs through Service Hubb AI — the follow-up text goes out automatically when a job closes, the thirty-day reminder fires on a schedule, the seasonal campaigns go to segmented lists. Once it's built, it runs without daily management. That's the point of having a CRM built for service businesses — the automations that matter are already built in, not something you have to configure from scratch. Retention through systems, not heroics.

The Referral Program: Turn Happy Customers Into a Sales Team

A satisfied customer who tells one neighbor about you is worth more than any ad you'll ever run. Word-of-mouth from a trusted source converts at a dramatically higher rate than any cold marketing channel — and it arrives pre-qualified because the person making the referral already pre-sold you.

Most service businesses get referrals passively. They do good work, hope customers mention them, and occasionally get a call that starts with “my neighbor recommended you.” That works to a degree. A referral program takes that passive flow and turns it into something intentional.

But before we get to the mechanics, understand why referrals actually happen. People don't refer your business because you offered them twenty-five dollars. They refer you because they want to be the person who has a “guy”. Everyone wants to be the one in their circle who can say “I know exactly who to call — let me give you his number.” That's social currency. When someone refers you they're extending their own reputation to vouch for yours. They're doing it because the experience you delivered made them confident enough to put their name behind you. That's the real referral engine — not the gift card, not the discount, not the incentive program. The experience.

That said, a well-timed incentive doesn't create referrals but it can accelerate them. It gives someone who was already going to mention you a specific reason to do it today rather than eventually. A discount on a future service sounds logical until you remember that most of your customers just got rid of everything they needed to get rid of. Offering them money off a future junk removal is like offering someone a discount on another meal right after they finished eating — the timing makes the incentive worthless. Instead offer something with immediate universal value. A $25 Amazon gift card sent digitally the moment their referral books a job costs you almost nothing relative to the lifetime value of a new customer, arrives instantly, and works for everyone regardless of whether they ever need your service again. Send it in the follow-up message. Remind them in the seasonal email. Mention it on the invoice. The incentive doesn't need to be large — it needs to be useful, and it needs to be communicated.

The most effective referral asks come in the moment right after a job goes well — when the customer is satisfied, the space looks great, and the goodwill is at its peak. "If you have any friends or neighbors who need this kind of help, we'd love the referral — and we'll take care of them the same way we took care of you." Simple, genuine, and timed right. That ask, made consistently after every good job, compounds into a steady referral stream over months and years.

But never confuse the accelerant for the engine. The reason people refer you is because you showed up, did the job right, and made them look good for recommending you. Build that first. The gift card is just a nudge.

Track where your new customers come from. If referrals are generating 20% of your new business, that number tells you the referral program is working. If it's 5%, you're either not asking or the ask isn't landing. Adjust accordingly.

Know Your Customers Well Enough to Surprise Them

The service businesses that build the deepest loyalty aren't just reliable — they're personal. They remember things. They treat returning customers differently than first-time callers. They make people feel like valued regulars rather than another ticket in the queue.

This doesn't require an elaborate operation. It requires a CRM with notes and the discipline to use it.

When a customer calls back, you should be able to pull up their record in thirty seconds and know what job you did for them, when, whether there were any issues, and any personal details that came up. "Hi Mrs. Lopez, good to hear from you — we did your garage cleanout back in April, right? How's it been holding up?" That thirty-second preparation turns a generic service call into a personal interaction. People remember how you made them feel long after they've forgotten the details of the job.

Tag your best customers. The ones who book repeatedly, refer others, and leave strong reviews are your VIPs. They deserve to know it. A handwritten thank-you card after a big job costs almost nothing. A small gift card to a local coffee shop for a customer who has referred three people costs $20 and generates disproportionate goodwill. An unexpected discount on a repeat booking — not asked for, just applied — surprises someone who expected to pay full price and creates a story they'll tell.

These gestures aren't expensive. They're memorable. And memorable is what gets you recommended.

Expand the Relationship Over Time

Every customer has more than one potential job in them. The homeowner who called for a one-time basement cleanout is also a candidate for seasonal junk pickups, dumpster rental for her bathroom renovation, and demo work when she finally tears out that old deck. She doesn't know that unless you tell her.

Make sure every customer knows your full range of services. Not a hard sell — a natural mention. "By the way, if you ever have a project where you need a dumpster, we rent those too. Most of our cleanout customers don't realize that until they're already calling someone else." One sentence. It plants the seed without being pushy.

Follow-up content can do this work passively. A simple monthly or quarterly email that highlights different services — what a dumpster rental looks like, a before-and-after from a commercial cleanout, a reminder about seasonal services — keeps your full capability visible without requiring a sales conversation.

The customer who starts with a $400 residential cleanout and over three years books two dumpster rentals, a deck removal, and refers two neighbors is worth several thousand dollars in lifetime revenue. That math changes how you think about every job. It's not a $400 transaction. It's the beginning of a relationship with a lifetime value that depends entirely on whether you stay in front of them.

Community as a Retention Tool

People support businesses they feel connected to. That connection can come from personal relationships — you remember their name, your crew is friendly, you follow up after the job. It can also come from community presence.

Community presence isn't just a brand-building move — it's a retention move. The customers who see your name on the little league jerseys, who watched your crew volunteer at the neighborhood cleanup, who noticed you sponsoring the charity event — those customers don't compare quotes the next time they need you. You're already the obvious call. Chapter 9 covers the specific tactics; the principle here is that the same community actions that build your brand also make existing customers less likely to leave.

When you do something in the community, tell people about it. Not to brag — to signal who you are. A post about the charity donation run you helped with, the little league team you sponsored, the neighborhood cleanup you participated in. These posts reach your existing customers and reinforce why they chose you. They also reach potential customers who are evaluating whether to call.

The businesses that become genuinely embedded in their communities — whose names come up naturally when neighbors ask for recommendations, whose trucks are recognized as belonging to a company that cares — build a moat that advertising cannot replicate.

The Compounding Effect

Customer retention works exactly like compound interest. You invest a small amount — a follow-up text, a referral ask, a seasonal email — and it pays a small return. That return generates a slightly larger customer base. The slightly larger base generates more referrals. The referrals generate more reviews. The reviews generate more first-time customers, some of whom you retain. Each cycle, the base grows.

The business that has been running a consistent retention system for three years has a fundamentally different foundation than the one still acquiring every customer cold. Lower cost per acquisition. Higher close rates. More predictable revenue. A review profile that compounds with every new review added. A referral network that grows because each retained customer represents a potential referral source.

And when a new competitor enters your market — better funded, slicker brand, aggressive on price — your retained customers don't move. They already have their person. They know your name. They have your cell number. That loyalty, built through consistent care over time, is the most durable competitive advantage a local service business can have.

The metric that tells you whether your retention system is actually working: your repeat and referral rate. What percentage of this month's jobs came from past customers or their referrals? Track it monthly. If it's growing, the system is compounding. If it's flat or declining, you have a problem even if total revenue looks fine — you're replacing customers you should be keeping. The full calculation and how it fits into your monthly business review is in Appendix A.

> *It's not your customer's job to remember you. It's your responsibility to make sure they never forget you.*

Go Deeper

For more on this and everything else in the book, the Haulers' Edge Newsletter goes deeper every week. Scan the code or subscribe free at HaulingHubb.com.

CH. 12 WHERE 80% OF YOUR MONEY IS ACTUALLY COMING FROM

> *"Doing less is not being lazy. Don't give in to a culture that values personal sacrifice over personal productivity." — Tim Ferriss*

Most service business owners work hard across everything equally. Every job gets the same energy. Every customer gets the same attention. Every service gets marketed the same way. Every lead gets chased with the same urgency.

And then they wonder why growth feels like pushing a boulder uphill.

The problem isn't effort. It's where the effort is going. Because in almost every service business, a small fraction of customers, services, and activities are generating the majority of revenue, profit, and growth — while everything else is just keeping you busy.

The 80/20 rule doesn't tell you to work less. It tells you to look accurately at where your results are actually coming from — and ruthlessly redirect your energy toward those things.

What Pareto Actually Means for Your Business

You've probably heard the 80/20 rule. The insight behind it isn't that the math is always exactly 80/20 — it's that results are almost never evenly distributed. A small number of inputs produce a disproportionate share of outputs.

In a service business that plays out across every dimension.

A handful of customers generate most of your revenue. One or two service lines deliver most of your profit. A couple of marketing channels produce most of your qualified leads. A small percentage of your time — the hours actually spent on high-value work — drives most of your real progress. Meanwhile, a small group of problem customers generate most of your headaches, disputes, and stress.

None of this is obvious until you look at the numbers. Most operators feel it intuitively — that some jobs are just better than others, that some customers are worth ten times what others are worth — but they never quantify it. When you quantify it, the imbalance becomes undeniable and the right decisions become obvious.

Run the Audit

You can't act on the 80/20 principle without data. Here's how to run the audit on your own business.

Revenue by service line. Pull your last twelve months of jobs and sort them by service type. Calculate total revenue and estimated gross margin for each category. The pattern will almost always show one or two service lines generating the majority of both. For a junk removal operation it might be large residential cleanouts and dumpster rentals carrying the revenue while single-item pickups eat crew time for thin margins. For a landscaper it might be installation and hardscape projects versus weekly maintenance. For a cleaning company it might be commercial contracts versus one-time residential jobs. Whatever the breakdown — find it. The service lines in the top 20% deserve more marketing attention, more capacity, and more pricing confidence. The ones in the bottom should be repriced, restructured, or cut as covered in Chapter 4.

Revenue by customer. Sort your customer list by total spend over the last year. In most service businesses you'll find that a small group of customers — contractors, property managers, real estate investors, high-frequency residential clients — are responsible for a disproportionate share of revenue. These are your VIPs. They deserve priority scheduling, personal relationship management, and proactive retention effort. Losing one of them is a material event for your business. Do you even know who they are right now?

Revenue by marketing channel. Track where your jobs are actually coming from. Google search, referrals, repeat customers, a realtor partnership, Nextdoor recommendations — they don't all produce equally. Most businesses find that two or three channels generate the overwhelming majority of quality leads while several others produce occasional work at high cost in time or money. The channels in your top 20% deserve more investment. The others deserve honest evaluation of whether they're worth continuing.

Time by activity. This one requires real honesty. Of the hours you spend each week, which ones directly produce revenue, relationships, or business growth? Which ones are administrative, reactive, or low-value? Answering phones for jobs you shouldn't be taking. Manually doing things a system could automate. Chasing a payment that's been outstanding for sixty days. Driving an hour each way for a job that barely covers costs. These aren't hypotheticals — they're the reality for most service businesses. The time you recover by eliminating or automating low-value activities is time you can put into the work that actually moves the needle.

What to Do With What You Find

The audit is only useful if it drives action. Here's what acting on it looks like.

Double down on your top 20%. The services delivering the best margins get more marketing emphasis, more capacity allocation, and more pricing confidence. The customers generating the most revenue get priority treatment, proactive follow-up, and the personal relationship management covered in Chapter 11. The marketing channels producing the best leads get more budget. You're not doing anything new here — you're doing more of what's already working.

Reprice or cut the bottom. The services barely breaking even need to either be repriced to a margin that justifies delivering them or removed from your lineup. The jobs that eat crew time for thin returns need a higher minimum charge or a stop altogether. This is the same principle from Chapter 4 applied to data instead of gut feel. Now you have the numbers to make the decision without second-guessing yourself.

Fire the problem customers. This one makes operators uncomfortable until they do it. The customers generating the most disputes, the most calls, the most invoice negotiations, and the most stress are almost never your highest-revenue accounts. They're the chronic late payers, the scope creepers, the people who need three calls before they book and then want a discount at the end. Letting them go frees time and mental energy that consistently exceeds the revenue they represent. We've fired many clients over the years — mostly by declining to work with them again when they won't agree to our standard terms and policies. The headaches from problem customers come in predictable forms: delayed payments, custom payment schedules that require constant chasing, scope that bleeds into adjacent jobs and disrupts routing, excessive calls that drag on and consume admin time. Every time we've let one of these accounts go, we've opened up capacity for better work. The calendar doesn't stay empty.

Automate the trivial many. The administrative tasks eating your time — follow-up texts, invoice reminders, review requests, scheduling confirmations — don't need you personally. They need a system. Every hour you recover from low-value administration is an hour available for high-value work. Chapter 21 covers the tools in detail. The principle applies here: use the 80/20 audit to identify which tasks to automate first.

The first move after running this audit: pick the single lowest-margin service that consumed more than 10% of your crew's time last quarter. Either raise its price by 20% this week or remove it from your website. Track what happens to your overall margin over the next 30 days. Almost every operator who runs this exercise is surprised — either the jobs disappear without revenue impact, or customers pay the higher price without pushback. Either outcome is better than the status quo.

The Revenue Conversation Most Operators Avoid

Here's what the 80/20 audit reveals that most operators don't want to see.

A significant portion of your current revenue is costing you more than you realize — in crew time, management attention, vehicle wear, and opportunity cost. Some of your busiest service lines are your least profitable when you account for True Job Cost. Some of your most frequent customers are your lowest-margin accounts.

Being busy is not the same as being profitable. A full schedule of the wrong jobs is worse than a half-full schedule of the right ones, because the wrong jobs consume capacity that could be filled with better work while generating the illusion of productivity.

The operators who focus on their best work outperform the ones who try to do everything — and the gap compounds over time. This isn't academic. It's the difference between the operator who grinds for ten years and wonders why growth stalled and the one who runs the same business with half the job volume and twice the profit because they stopped doing the wrong things.

The hardest part of the 80/20 mindset isn't the analysis. It's the permission to stop doing things that generate revenue but not profit. There's a psychological pull toward busyness — toward full trucks and ringing phones — that makes it feel irresponsible to turn down work. That pull is the enemy of efficiency.

Revenue is not the goal. Profit is. And profit comes from doing more of the right things, not more of everything.

Make It a Habit

The 80/20 audit isn't a one-time exercise. Markets shift. Your service mix evolves. A customer type that was a top performer a year ago might not be today. A marketing channel that was generating consistent leads six months ago might have cooled.

Run the audit quarterly at minimum. It takes less than an hour when your books are organized. Pull the numbers, sort by the key dimensions, identify what's changed, and adjust accordingly. The operators who build this habit into their quarterly routine make better decisions faster because they're always working from current data instead of assumptions that have gone stale.

The questions to ask every quarter: What's my actual top 20% right now? Has anything that used to be in the top 20% slipped? Is anything new breaking through that deserves more attention? What in the bottom 20% am I still holding onto that I should cut?

Those four questions, answered honestly with real numbers four times a year, will do more for your business than most of the other strategic thinking you'll do.

Work on the right things. The results compound. Everything else is just motion.

The scorecard and monthly review system — every metric, how to calculate it, and what to do with the result — is in Appendix A at the back of the book. Run the 80/20 audit in this chapter to find your highest-leverage opportunities, then use the scorecard to track whether your actions are actually moving the numbers.

Go Deeper

For more on this and everything else in the book, the Haulers' Edge Newsletter goes deeper every week. Scan the code or subscribe free at HaulingHubb.com.

PART 3: MARKETING & LEAD FLOW

CH. 13 YOUR WEBSITE IS EITHER SELLING OR SLEEPING

"Always deliver more than expected." — *Larry Page*

Your website is working right now. The question is whether it's working for you or against you.

Every hour your site is live, potential customers are landing on it, forming an impression in five seconds or less, and deciding whether to call or keep scrolling. If your site is slow, confusing, hard to read on a phone, or missing a clear way to contact you — they're gone. Not to think it over. To your competitor who made it easier.

Most service business websites are sleeping. They exist. They have a phone number somewhere. They list some services. And they convert a fraction of the traffic they should because nobody ever treated them like a sales tool.

This chapter fixes that. If you already have a site and think it's performing fine, run this check first. Five indicators that tell you whether your site is actually converting or just existing: Is your phone number in the header with click-to-call enabled on mobile? Does each primary service have its own dedicated page? Does your site load in under three seconds on a phone — test it free at PageSpeed Insights? Is your Google review rating and count visible without the visitor having to scroll? And do you know your current conversion rate — what percentage of site visitors contact you? If you can answer yes to all five, your site is working. If any are missing, this chapter shows you exactly what to fix and why each one costs you leads.

Two Audiences, One Website

Your website has two audiences simultaneously — the human visitor and the algorithm deciding whether to show your business in search results.

The human visitor needs clarity and speed. They want to know in five seconds what you do, where you do it, whether you're trustworthy, and how to reach you. Friction at any step loses them.

The algorithm — Google's search engine and the AI tools increasingly influencing which businesses get recommended — needs structured, consistent, trustworthy data. Your services, service area, pricing structure, reviews, and business details. When someone searches "best junk removal in [your city]" and Google assembles its local results, it's pulling from everything that exists about you online. Incomplete or inconsistent information pushes you down.

These two audiences aren't in conflict. A website built to convert humans is almost always built in a way that algorithms can read. You just need to hold both in mind.

Content: Connect, Convert, and Get Found

Content is what your website says and how it says it. It does three jobs: connects with the visitor by showing you understand their problem, converts them by making the next step obvious, and gets found by using the language people actually search.

Your headline is the first five seconds. Most service business homepages open with something like "Welcome to Joe's Junk Removal — Serving the Greater Metro Area Since 2011." That tells the visitor nothing useful and gives them no reason to stay. Lead with what they actually want to know that answers their question immediately. "Same-Day Junk Removal in [Your

City] — We Handle Everything, You Just Point." Specific, action-oriented, immediately confirms they're in the right place.

Follow the headline with a single obvious call to action. Get a Quote. Check Availability. Book Now. One button, visible without scrolling, every time someone lands on your page.

Dedicated pages for each service. Don't bury all your services on one generic page. A homeowner looking for a garage clean-out and a contractor looking for dumpster rentals have different needs, different questions, and different search terms. Give each service its own page — what's included, what you won't take, pricing ranges, turnaround time, and a call to action. "Dumpster Rental in [Your City]" as a standalone page ranks significantly better than a generic services page that mentions dumpsters in a list. It also gives algorithms a clean structured answer to a specific search query.

Answer questions before they're asked. The most common objections in a service business are almost always the same. How much does it cost? Are you licensed and insured? What won't you take? How does scheduling work? A simple FAQ section on your website removes friction for the human visitor and creates machine-readable question-and-answer content that search tools frequently pull directly into results. A question like "Are you licensed and insured in [City/State]?" answered with a direct yes and your credentials is exactly the kind of structured content that feeds local search visibility.

Publishing honest pricing on your website does something most operators don't anticipate — it filters your leads before the phone rings. The customer who reads your pricing page and still calls is already comfortable with your rate. The one who leaves because the price is too high was never going to be a good customer anyway. Transparency in your content doesn't cost you jobs. It costs you the jobs you didn't want.

Answer what your competitors refuse to. The five topics

buyers research most before hiring any service business are always the same — cost and pricing, problems and negatives, comparisons to alternatives, reviews and validation, and best-of content. Most operators avoid publishing honest answers to all five because it feels uncomfortable. That's exactly why you should. A competitor who answers what everyone else refuses to address captures the customer before the phone ever rings — and in AI search, the business with the most direct honest answers is the one that gets cited.

Build a page for each of these five topics. Keep them honest, specific, and written in plain language. These five pages will outperform any generic service page you've ever published because they answer what your buyer is actually asking.

Real social proof throughout, not just on a testimonials page. Real reviews from real customers with names and locations build more trust than any marketing copy you'll write. Sprinkle them across your site — on the homepage, on service pages, near your contact form. Your website should reinforce what your Google profile shows — and as covered in Chapter 9, most customers are reading both before they call. Don't treat it as a separate destination.

Blog content — local and specific only. A few years ago, writing general informational articles — "how to dispose of a mattress" or "10 tips to prepare for junk removal day" — built meaningful organic traffic. That's changed. AI Overviews now answer most generic informational queries before anyone clicks a website, and blog traffic for informational content has declined sharply across virtually every industry. What still works is locally-specific content — "where to dispose of old electronics in [your county]," "what goes in a dumpster in [your state]," content tied to a specific place that a generic AI answer can't fully replace. Keep your content efforts focused there. Two or three genuinely locally-relevant articles a year is more valuable than ten generic ones.

Design: Remove Every Reason to Leave

Design isn't about looking impressive. It's about removing friction. Every extra click, every confusing layout, every piece of information that's hard to find is a reason for the visitor to leave and call someone else.

Speed is non-negotiable. More than half of mobile visitors abandon a page that takes more than three seconds to load *(Google, 2018)*. Compress your images. Use a reliable host. Remove plugins or scripts you don't need. Run your site through Google's PageSpeed Insights and fix what it tells you. A slow website isn't just a bad experience — it directly hurts your search ranking because Google treats page speed as a ranking signal.

Mobile first, always. The majority of local service searches happen on a phone. If your site requires pinching to zoom, has buttons too small to tap, or buries your phone number three scrolls down, you're losing the majority of your inbound traffic before they can even decide whether to call. Test your site on your own phone right now. If anything frustrates you, it's frustrating your customers. Google indexes the mobile version of your site first — if your mobile experience is weak, your rankings follow.

Navigation should be obvious. Home, Services, Reviews, Contact. That's the path most visitors take. Make it visible, simple, and consistent across every page. Your phone number belongs in the header of every page with click-to-call enabled on mobile. A visitor who has to hunt for your contact information will stop hunting and open the next result.

Real photos outperform stock photos every time. A before-and-after of a real garage cleanout your crew did last week does more for trust than any staged stock image. Real jobs, real trucks, real crew members. Keep file sizes small so they don't slow the page, and add descriptive alt text to every image — this helps accessibility and gives search engines structured information about what your business actually does.

Then go one step further. Put names under the faces. A crew photo with "Left to right: Marcus, Deon, and Tyler — your team." does more for trust than any marketing copy on the page. People hire people. Give them someone to hire.

Video proves what photos can only suggest. Photos show the result. Video shows the people, the process, and the personality behind it — and that combination builds trust faster than any other medium available to a local service business. A two-minute video filmed on your phone showing a real crew doing a real job does more for your conversion rate than a professionally produced brand spot.

For us, the clearest example is the Grizzly Bag — a dumpster bag product that many customers hadn't seen before. When a caller expressed interest but wasn't sure what they were getting, we'd send them to the service page with the walkthrough video. That video did the closing. Customers who had questions going into the call came out of it ready to book because they'd seen the product in action, watched how the process worked, and understood exactly what they were paying for. The video replaced fifteen minutes of explanation with two minutes of proof.

Film your next job with the customer's permission. Show the process from arrival to completion. Post it to your Google Business Profile, your website, and your social media. The production quality matters far less than the authenticity — customers can feel the difference between a real moment and a staged one, and real moments are what get you hired.

Some Good News About AI and Local Search

Before we get into the how, let's address something you've probably been hearing about: AI search, Google AI Overviews, and whether the whole game is changing under your feet.

Here's what the data actually shows as of early 2026: local service businesses are among the least disrupted categories in the entire search landscape. While AI Overviews have expanded aggressively across informational and educational content — recipe sites, news publishers, how-to articles — local search queries have remained largely protected. Research tracking millions of keywords found that AI Overviews appeared on fewer than 0.01% of local search queries by late 2025, down from an already small 0.14% earlier in the year. Local searches have the smallest share of AI Overview appearances of any major search category, largely because Google already has highly effective tools for local intent — the map pack, Local Services Ads, and the Google Business Profile — and doesn't need to replace them with AI-generated summaries.

That's genuinely good news for a junk removal operator, a landscaper, a plumber, or a cleaning company. The person searching "junk removal near me" or "dumpster rental in [your city]" is not getting an AI-generated answer instead of local business results. They're seeing a map pack, business profiles, and paid ads. The fundamentals that have always driven local search visibility — a strong GBP, consistent reviews, a fast mobile website, local keyword optimization — are exactly what work in the current environment. Nothing about AI search changes that. If anything the fundamentals matter more because they're also what feeds the AI-powered local packs that are slowly emerging alongside traditional results.

There is one new development worth knowing. Google has launched AI Mode — a Gemini-powered search experience that functions more like ChatGPT, generating comprehensive answers with citations but without the traditional list of search results. It's currently opt-in and represents a small fraction of search behavior. As it grows it will matter more, and the businesses that get cited in AI Mode answers will be the ones with the strongest overall digital presence — authoritative GBP, consistent reviews, well-structured website content. Same fundamentals, new delivery mechanism. We cover AI search in full in Chapter 34.

The bottom line going into this chapter: stop worrying about AI breaking local search and start building the website and digital presence that wins local search the way it's always been won — with clarity, speed, relevance, and trust.

Local SEO: Getting Found Before the Click

The best website in the world produces nothing if nobody finds it. For a local service business, SEO doesn't need to be complicated. The fundamentals are consistent and learnable — and as we established at the start of this chapter, they're exactly what works in the current AI search environment.

Use the language your customers actually search. They type "junk removal in Denver" or "dumpster rental near me" — not "waste management solutions in the metropolitan area." Use specific service-plus-city phrases in your page titles, headlines, and content. A service page titled "Junk Removal in Atlanta, GA" will rank for that term if the rest of the page backs it up. Create a separate page for each major service and each primary city or town you serve. Don't duplicate — customize each page with something specific to that location.

Your Google Business Profile is your most important local asset. It feeds the map pack, Google Maps, Local Services Ads, and the AI-powered local packs beginning to appear on mobile. Keep it complete and current. Your name, address, and phone number must match your website exactly across every platform. Every inconsistency creates ambiguity that pushes you down in results. Upload real photos regularly — jobs, trucks, crew. Respond to every review. Post updates when you have something worth saying. An actively maintained profile ranks better than a dormant one every time. Full GBP strategy was covered in Chapter 9. Apply all of it here.

One important note on GBP expectations: direct call volume from Google Business Profiles on mobile has been declining as Google redesigns how local results display. The solution isn't to abandon GBP optimization — it's to use your GBP to drive clicks to your website rather than relying solely on the call button. A strong GBP that drives website visits converts better than one optimized only for direct calls.

On-page basics matter more than most small business sites reflect. Each page needs a unique title tag — about 60 characters, service plus location — and a meta description of 150–160 characters. Use clear heading structure: H1 for the page title, H2 for major sections. Write descriptive URLs — /services/junk-removal not /page?id=47. These aren't complicated moves. They're the baseline most small business sites are missing, and fixing them produces measurable ranking improvements.

Structured data markup gives algorithms explicit information. Schema markup is a small amount of code that tells search engines and AI tools exactly what your business is, what it does, where it operates, and what customers say about it. At minimum, implement Local Business schema with your address, phone, and hours. Add Service schema for each service page. Add FAQ schema for your question-and-answer sections. Add Review schema to display ratings. This is what powers rich results and increasingly feeds AI-generated recommendations. It's not complicated to add — your web developer can implement it in an hour or your website platform may support it natively.

Build local citations and links. Get listed on reputable directories — Yelp, HomeAdvisor, Angi, your local Chamber of Commerce. Consistent presence across these platforms reinforces your legitimacy. When you sponsor a local event, partner with a charity, or join a trade association, ask for a mention on their website. Each external link from a credible local source adds authority that compounds slowly but meaningfully.

Reviews are a ranking factor. Google uses review quantity, recency, and rating as signals in local search ranking. The review generation system from Chapter 10 isn't just a reputation tool — it's an SEO tool. More recent reviews from real customers in your market improve your position in the local pack. Review text that mentions your city and service type adds keyword relevance. Respond to every review — the responses contain keywords too.

Local Services Ads — know they exist. This isn't a Google Ads chapter — that's Chapter 14 — but Local Services Ads deserve a mention here because they've expanded dramatically. As of early 2026, LSAs appear on roughly 22% of local service queries on mobile, up from 11% at the start of 2025. They're a pay-per-lead format specifically designed for local service businesses, they appear above standard Google Ads in search results, and they carry a Google Verified badge that builds immediate trust. If you're investing in paid local visibility, LSAs are the first conversation to have. Chapter 14 covers the full paid search picture including LSAs.

Converting Visitors Into Calls and Bookings

Traffic without conversion is just numbers. Here's what actually converts.

Make contact as easy as possible across every method. Some people call. Some fill out forms. Some send texts. Some want to chat. Give them every option and make all of them obvious. Phone number in the header with click-to-call on mobile. A short contact form — name, phone or email, service needed, nothing else. A text option where customers can send photos of the job. Live chat or a chatbot for instant questions. Every additional contact method you offer is another visitor you catch who wouldn't have used a different channel.

Frictionless booking converts better than everything else. The moment a customer decides they want to book, the business that makes it easiest to lock in a time wins. An online booking option — even a simple "Request a Pickup" form that asks for a preferred date and lets you confirm — creates a sense of progress that keeps customers from shopping around overnight. Reducing the gap between "I want to book" and "I'm booked" directly improves your conversion rate.

Trust signals near the call to action. Licenses, insurance, satisfaction guarantees — place these near your quote form and contact button, not buried in an about page. A visitor about to let a service company onto their property wants confirmation they're making a safe choice. The trust signal at the point of decision makes the difference between submitting and second-guessing.

Capture leads who aren't ready to book. Not every visitor is ready to commit on the first visit. A free estimate offer, a small discount for first-time customers, or a helpful local resource — "What can and can't go in a dumpster in [your state]" — captures contact information so you can follow up. The lead you capture today who books next week is worth capturing.

Automation: The Website Works While You Sleep

A website with the right automation captures leads, responds to inquiries, and follows up with prospects around the clock without you touching a phone at 10pm.

An after-hours autoresponder on your contact form and chat costs nothing to set up and converts leads that would otherwise go cold overnight. "Thanks for reaching out — we received your message and will be in touch first thing tomorrow morning. Text us photos of the job here for a faster quote." That one message shows professionalism, buys you time, and prevents the customer from calling your competitor before you respond.

A CRM that captures every inquiry — web form, phone call, text, chat — and tracks it through your pipeline is the backbone of lead management at any scale. At Grizzly Junk Pros we run Service Hubb AI, which handles web lead capture, pipeline tracking, automated follow-up sequences, and review requests in one system — all built specifically for service businesses so you're not adapting a generic CRM to fit your operation. Without a CRM, leads fall through the cracks. With one, every lead gets followed up regardless of how busy you are.

Automated follow-up sequences for leads that don't immediately book — a next-day reminder, a three-day check-in, a week-later message — recover a meaningful percentage of people who expressed interest and went quiet. Most of them aren't gone. They're just busy. A simple automated nudge brings them back at a fraction of the cost of acquiring a new lead.

Start Simple, Build From There

If you're starting from scratch or significantly behind, here's the order of operations.

Get a clean, fast, mobile-optimized five-page site live first — home, services, about, reviews, contact. Real photos, clear headline, click-to-call, short form. That alone outperforms most of your competitors. Then add dedicated service pages with local SEO. Then FAQ content and schema markup. Then chat and online booking. Then automation and CRM integration. Each layer adds conversion and visibility. None of it requires a developer or a large budget — it requires sequenced effort applied consistently over time.

The website is never finished. Use Google Analytics to track what visitors do — where they land, what they click, where they leave. Small improvements made consistently compound into significant results over months and years.

The local search landscape is shifting. But the fundamentals that have always driven local visibility — a fast clean website, an optimized GBP, consistent reviews, structured content, and frictionless conversion — are exactly what work today and exactly what most of your competitors still haven't gotten right.

> *Your website is either working for you right now or it isn't. Treat it like your best employee — keep it sharp, keep it current, and it will keep producing.*

Go Deeper

For more on this and everything else in the book, the Haulers' Edge Newsletter goes deeper every week. Scan the code or subscribe free at HaulingHubb.com.

CH. 14 OWN YOUR ZIP CODE ON GOOGLE

> *"Without data, you're just another person with an opinion." — W. Edwards Deming*

Most service business owners treat Google Ads one of two ways. They set up a campaign, let it run, and check it occasionally when they remember. Or they throw money at it for a few months, don't track results carefully, decide it doesn't work, and quit.

Both approaches burn budget and produce nothing useful.

Google Ads works. For local service businesses it's one of the most direct paths from ad spend to booked job that exists. Someone types "junk removal near me" — they have a problem right now and they want it solved. Your ad shows up. They call. That's a buyer who raised their hand. You didn't have to build brand awareness or nurture them through a funnel. You just had to be visible at the right moment.

But visible on Google isn't enough. Visible with the right structure, the right message, the right budget controls, and the right follow-through is what converts that visibility into revenue.

How Google Ads Actually Work for Local Service Businesses

Before getting into mechanics, you need to understand what the paid search landscape actually looks like — because it keeps changing, and many businesses are running campaigns built for the version that existed two years ago.

There are now effectively three paid layers on Google for local service businesses and they stack from top to bottom on the results page.

Local Services Ads sit at the very top. These are the listings with the blue Google Verified badge — formerly called Google Guaranteed — that appear above everything else including standard Google Ads. They run on a pay-per-lead model rather than pay-per-click, require background checks and license verification, and are deeply integrated with your Google Business Profile. As of October 20, 2025, Google consolidated all its trust badges into a single Google Verified badge, discontinuing the Google Guaranteed designation and its associated money-back guarantee (Coalmarch, 2025). LSAs have expanded dramatically — as covered in Chapter 13, they now appear on roughly 22% of local mobile queries. For service businesses that qualify, LSAs are the first paid conversation to have. We cover them in their own section below.

Standard Google Search Ads sit below LSAs. These are the traditional pay-per-click campaigns where you bid on keywords, write ad copy, and pay when someone clicks through to your site or calls. This is what most people mean when they say "Google Ads" and it's the primary focus of this chapter.

Your Google Business Profile and organic results sit below both. We covered those in Chapters 9 and 13. The paid layers above them make the organic presence more important, not less — a prospect who sees your ad, then sees your GBP with 150 reviews in the map pack, then sees your website in organic results, has encountered your name three times before making a decision. That reinforcement converts.

Understanding all three layers and how they work together is how you own your zip code. Let's build them.

Local Services Ads: Start Here

If you haven't set up Local Services Ads yet, do it before you spend another dollar on standard Google Ads.

LSAs appear above standard ads in search results. They show your business name, review rating, general location, and the Google Verified badge. They're pay-per-lead — you pay only when a prospect calls or messages through the ad, not when someone simply clicks. For most local service businesses the cost per lead through LSAs is competitive with or better than standard search campaigns, with the added advantage of position priority.

To run LSAs you need a verified Google Business Profile with accurate information — GBP verification is now mandatory for all LSA advertisers, and all customer reviews flow through your GBP rather than a separate LSA profile. You also need to pass Google's background check and license verification process. The setup takes a few weeks and requires documentation — business license, insurance certificates, and owner background check. It's an investment of time upfront that pays back in priority placement.

A few things to know about how LSAs work. Google's lead dispute and credit system changes periodically — check the current LSA interface for the latest mechanism. The principle that remains stable: submit feedback on every lead you receive, whether it was a good lead or not. Google uses that data to refine its algorithm and may issue credits based on consistent feedback patterns.

Track message leads and phone call leads separately — Google provides cost reporting for each, and they convert at different rates. Most operations find phone leads generate higher-quality conversations. Track both and allocate toward whichever delivers better results for your market.

Your LSA ranking is driven by review rating, review recency, responsiveness, and proximity. The review generation system from Chapter 10 isn't just a reputation tool — it's a direct LSA ranking factor. A business with 200 recent reviews ranks higher in LSAs than one with 40 stale ones. Reviews now feed both your GBP and your LSA profile simultaneously.

Google Search Ads: The Core Campaign System

With LSAs running at the top, your standard search campaigns handle the volume and targeting precision below them.

Structure by service, not by budget. The most common mistake is throwing all your services into one campaign. Create a separate campaign for each major profit center — Junk Removal, Dumpster Rental, Demolition, Estate Cleanouts. This lets you match every keyword and ad to exactly what the customer searched. Someone looking for "dumpster rental for renovation" sees a dumpster ad. Someone looking for "garage cleanout service" sees a cleanout ad. The precision improves relevance, raises Quality Scores, lowers your cost per click, and makes budget allocation by service profitability possible. A $50 CPA on a $600 demo job makes sense. The same CPA on a $150 single-item pickup doesn't. Keep them separate so you can see that math clearly.

Inside each campaign, break ad groups by job type. A Junk Removal campaign might have separate ad groups for Appliance Removal, Estate Cleanouts, and Construction Debris. Each ad group contains a small cluster of closely related keywords and ads written to match them specifically. "Fast Appliance Removal — Same Day Available" is a more compelling response to an appliance removal search than a generic "We Remove All Junk" headline. Narrow ad groups produce better Quality Scores, better click-through rates, and ultimately lower costs.

Use keyword match types with intention.

Exact match — [junk removal near me] — gives you maximum control and laser precision. Lower volume but highest intent.

Use it for your core money terms.

Phrase match — "junk removal" — is your workhorse. Shows for searches containing your phrase with additional context words. Captures "same-day junk removal service" and "junk removal for contractors" without going dangerously broad.

Broad match — junk removal — can surface new opportunities but burns budget on irrelevant searches without strong management. If you use it, pair it with a robust negative keyword list and weekly search term reviews. Broad match under close management can work. Broad match set and forgotten is a budget leak.

Negative keywords are as important as positive ones. Build a universal negative list applied to every campaign: free, jobs, hiring, salary, DIY, cheap, how-to, truck for sale, how to start a junk removal business, and any cities or regions you don't serve. Run a search terms report weekly and add new negatives constantly. A single negative keyword that blocks irrelevant clicks can save hundreds of dollars a month. This is where most operators leave money on the table — they focus on what to bid on and ignore what to block.

Ad Copy That Converts

Your ad is often the first impression a potential customer has of your business. It needs to answer three questions instantly: do you do what I need, are you local, and why should I call you instead of the next result.

Lead with the specific service and location. "Same-Day Junk Removal in [City] — Licensed & Insured Crew" tells the searcher immediately that you're relevant and local. Generic headlines like "Best Junk Removal Service Available Now" tell them nothing they couldn't read on every other ad.

Use your differentiators as headlines, not afterthoughts. If you offer same-day service, make it the headline. If you've done 500 jobs in their city, say it. If you have a 4.9 star average, put it in the ad. These specifics do more work than any clever phrasing.

Responsive Search Ads — the current standard Google Ad format — let you provide up to 15 headlines and four descriptions that Google rotates to find the best combinations. Keep at least two to three RSAs per ad group and test a new headline variation monthly. Small wording changes occasionally produce significant CTR improvements and you won't find them without testing.

Extensions — now called Assets in Google's updated terminology — make your ad larger on the page and more useful before anyone clicks. Enable every relevant one: Location Asset to prove you're local and get into Maps, Call Asset so mobile users can tap to call without clicking through, Sitelinks to send people directly to pricing or booking pages, Callouts to highlight trust points like "Licensed & Insured" and "Free Estimates," and Structured Snippets to list your services. These cost nothing extra and each one pushes competitors further down the page.

Bidding and Budget: Math First

Start with what a job is worth to you. If an average job generates $300 revenue and $150 profit, how much of that profit are you willing to spend to acquire it? That's your target cost per acquisition. If you'd spend $40 to land a job netting $150, your CPA target is $40. If your close rate on leads is 50%, you need $80 in ad spend to generate each job — meaning you need to budget $80 per acquired job, not $40 per lead. Run the math before you set budgets, not after.

Allocate budget by campaign based on service margins. High-margin services like demolition deserve more budget than single-item pickups. Google splits spend evenly by default. Override that default and direct spend toward your most profitable service lines.

Manual versus automated bidding — know when to use each.

Manual bidding gives you direct control over what you pay per click on each keyword. It works well when you're gathering early data, when you have a small number of high-value keywords to manage, and when you're detail-oriented enough to review bids weekly. The cost is time — manual bidding requires consistent attention to avoid overpaying on some terms and underpaying on others.

Automated Smart Bidding — Target CPA, Maximize Conversions, Target ROAS — lets Google's AI adjust bids in real time based on device, location, time, user behavior signals, and historical conversion data. In 2025 AI became fundamental to how Google Ads operates rather than an optional feature, and Smart Bidding now processes signals humans genuinely cannot monitor manually. Smart Bidding works well once you have conversion data — typically 30 or more conversions in the past 30 days gives the algorithm enough to work with. Before that threshold, manual or a hybrid approach produces more predictable results.

A practical approach: start manual to accumulate 30 or more conversions, then transition to Target CPA Smart Bidding with your calculated CPA target. Monitor performance for two to three weeks after the switch before judging results — the algorithm needs a learning period.

For a service business in a mid-sized market just starting out, expect to spend $30–$60 per day minimum to generate enough click volume for meaningful data within 30–60 days. Below $20/day, results are too sparse to optimize from. If your market is smaller or more competitive, adjust that floor accordingly — check what your target keywords cost per click and back into a daily budget that generates at least 5–10 clicks per day.

If your budget makes reaching 30 conversions in a month impractical, skip Smart Bidding entirely and stay on manual CPC indefinitely. LSAs are a better primary investment at small scale — pay per lead rather than per click, and the verification badge carries trust signals Google Ads alone doesn't provide.

Run ads when customers search and you can respond.

Service businesses get calls. If your team can't answer calls after 8pm, pausing ads at 8pm saves real money. A click that goes to voicemail is a lead you paid for and lost. Use ad scheduling to align your ad exposure with your actual availability. Start broad — say 6am to 9pm — and use your search term and conversion data to refine over time. If Saturday mornings consistently produce jobs at half your average CPA, increase bids on Saturday mornings. Let the data tell you when to be aggressive.

Landing Pages: Where Clicks Become Calls

You paid for the click. Now the landing page either converts it or wastes it.

Message match is the most important principle. The landing page must deliver exactly what the ad promised. If your ad says "Same-Day Junk Removal — Call Now," the page headline should confirm same-day availability and make calling effortless. If the keyword was "dumpster rental prices," the page should show dumpster pricing prominently. When the ad and page say the same thing, conversion rates climb. When they don't, the visitor bounces and you've wasted the click.

Create dedicated landing pages for your primary ad groups rather than sending everyone to your homepage. A visitor searching "estate cleanout service" who lands on a page specifically about estate cleanouts — with relevant photos, relevant copy, and a relevant call to action — converts at a significantly higher rate than one who lands on a generic services page and has to figure out whether you handle their need.

Speed determines whether the page gets a chance. Every extra second of load time cuts conversions meaningfully. Target under three seconds on mobile. Compress images, keep the layout simple, test with PageSpeed Insights. Most of your searchers are on phones. A page that takes five seconds to load on a phone loses a significant share of those visitors before they've seen a single word.

One clear call to action. Not five options — one. If you want calls, display the number large with tap-to-call on mobile and nothing competing for attention. If you want form submissions, put the form above the fold with three fields maximum. Add a trust signal near the CTA — a review count, an insurance badge, a satisfaction guarantee. The trust signal at the moment of decision reduces hesitation.

Tracking: If You're Not Measuring It, You're Guessing

Most service business owners who say Google Ads doesn't work aren't tracking calls. They're running ads, the phone is ringing, and they have no visibility into which keywords or ads produced those calls so they assume nothing is working.

Set up call tracking before you spend a dollar. Google's built-in call forwarding with call extensions tracks calls generated directly through your ad. For deeper tracking — which specific keyword drove each call, recorded calls, attribution back into your CRM — a tool like CallRail runs $30–$100 per month and pays for itself with a single recovered job per month. Without call tracking you're making optimization decisions without data, which means you're optimizing on guesswork.

Also verify that form submissions, chat conversations, and online bookings are firing as conversion events in Google Ads. Every contact method that isn't tracked is a conversion hole — your optimization decisions will be skewed toward whatever you are measuring.

The Weekly and Monthly Rhythm

Google Ads is not a set-and-forget system. It requires consistent attention at two cadences.

Weekly — 20 to 30 minutes. Check click-through rate, conversion rate, cost per conversion, and top-spending keywords. Run the search terms report and add negatives. Pause any keyword that has spent significantly with zero conversions. Flag anything that looks wrong — sudden budget spikes, conversion tracking breakdowns, ad disapprovals.

For service businesses running their own ads, target a click-through rate of 5–10% on high-intent searches. Conversion rate of 5–15% on quality traffic. Cost per lead at or below your CPA target. If any of those are significantly off, investigate immediately — the cause is almost always a search terms problem, a landing page problem, or a tracking problem.

Monthly — deeper review. Pause keywords that consistently exceed your CPA target. Shift budget from underperforming campaigns to proven ones. Review ad performance and pause losing variations. Write new challenger headlines. Check landing page bounce rates. Audit your negative keyword list. Write a short note summarizing spend, conversions, CPA, what changed, and what to test next month. That log becomes your optimization history and prevents you from making the same mistake twice.

The rhythm is what separates the operators who build a consistently improving lead machine from the ones who blow budget and conclude Google Ads doesn't work.

When to Get Help

Running Google Ads while operating your business is demanding. There's a point where the time cost of managing campaigns yourself exceeds what you'd pay a competent professional to do it. That point is different for everyone but honest self-assessment is worth doing.

Consider outside help when you routinely go weeks without logging in, when performance has plateaued and you've exhausted what you know, when you're scaling spend and small inefficiencies are getting expensive, or when campaign management has become a source of consistent stress.

When evaluating agencies or freelancers, require that all campaigns run in your account billed to your card — not theirs. Insist on full account access at all times. Walk away from anyone who can't explain their strategy in plain language, promises unrealistic results, or produces reports that never show anything changing. A good paid search partner reports clearly, optimizes consistently, and shows you exactly where your money went and what it produced. Red flags: set-and-forget mentality, opaque reporting, no discovery process, and account ownership that lives in their infrastructure rather than yours.

At Adimize we manage paid search for service businesses and the approach is the same whether you're running it yourself or working with a partner — structure by service, track everything, optimize weekly, and let the data drive decisions.

Putting It Together

LSAs at the top of the page. Standard search ads below them. GBP and organic reinforcing both. Call tracking connecting everything to real revenue data. Weekly pulse checks and monthly deep dives keeping the system sharp.

That's what owning your zip code on Google actually looks like. Not just showing up somewhere on the page — being the visible, credible, obvious choice at the moment someone in your market decides they need what you do.

The operators who build this system and maintain it consistently don't need to wonder where their next job is coming from. They already know.

> *The leads are already out there. Make sure you're the one they find.*

Go Deeper

For more on this and everything else in the book, the Haulers' Edge Newsletter goes deeper every week. Scan the code or subscribe free at HaulingHubb.com.

CH. 15 EVERY JOB SITE IS A BILLBOARD

"Marketing is not an event, but a process. It has a beginning, a middle, but never an end." —Jay Conrad Levinson

Almost every experienced operator I know thinks they've outgrown field marketing. Most of them are leaving a meaningful number of leads on the table every week because they stopped doing the basics or never built them into a system.

The door hangers, yard signs, and truck visibility tactics in this chapter aren't amateur-hour moves. They're the highest-ROI local marketing available to a service business — when they're systematized, tracked, and built into the crew's job completion routine rather than done ad hoc when someone remembers. A three-truck operation with a consistent five-around program, tracked QR codes, and crew-deployed yard signs is running a local marketing system that generates measurable inbound at near-zero cost. That's not what you did when you were broke. That's what the best operators in mature markets still do because it works.

The best marketing you'll ever do doesn't feel like marketing to the person receiving it.

It's your truck parked on a busy corner. It's the door hanger a homeowner finds the same day they watched your crew work next door. It's a yard sign that makes three neighbors ask the same question before you've driven off the block. It's a sticky note that ends up on someone's fridge for six months until the day they finally clean out the garage.

None of this requires a big budget. All of it requires showing up consistently and making every job site, every truck, and every finished job do double duty — the work and the marketing happen at the same time.

This is the marketing most service businesses overlook because it's not digital and it doesn't have a dashboard. But for local service businesses competing in specific neighborhoods, it's some of the highest-ROI activity you can do. Here's how to build it into your operation.

Your Truck Is Already a Billboard

Every wrapped, branded truck you operate is generating impressions every day it's on the road — estimates suggest a vehicle moving through a typical metro area generates 30,000 to 70,000 visual impressions daily *(Outdoor Advertising Association of America)*. Once you've paid for the wrap, those impressions cost nothing.

Most operators underutilize this. The truck gets seen on the way to and from jobs. Fine. But a truck parked strategically — near a busy intersection, outside a hardware store, at a subdivision entrance — keeps generating impressions when it's sitting still. A full day on a visible corner while you work a nearby job is a full day of passive exposure to exactly the homeowners you want to reach.

Brian Scudamore of 1-800-GOT-JUNK? famously grew his early business largely on the back of a single well-placed branded truck. The phone number was massive. The truck was visible. It ran constantly through neighborhoods. People saw it repeatedly, the name stuck, and when they needed junk removed, they called the number they'd seen a dozen times. That's the compounding effect of consistent visual presence.

A few practical rules. Keep the branding legible from a moving car — company name, phone number, website. Large enough to read in under two seconds at 30 miles per hour. Move the truck regularly so it doesn't look abandoned. Drive it courteously — one incident in a branded truck creates a negative brand association that no amount of advertising undoes.

Every dumpster you rent is doing the same thing. A 20-yard container sitting in someone's driveway for a week is a week of neighborhood-level advertising. Make sure your name and number are on it prominently. Every neighbor who drives or walks past that dumpster is a potential future customer seeing your brand.

The Five-Around Rule

Every job you complete is an opportunity to generate the next one before you leave the block.

The five-around rule is simple — when you finish a job, hit the five closest houses with a door hanger. Not fifty houses. Five. The ones who could see your truck from their driveway, who may have watched your crew work, who potentially talked to your customer while you were loading. Those neighbors already have social proof working in your favor without you saying a word.

Door hangers work because of the context they arrive in. A homeowner finds it on their door the same day they watched a professional crew work next door for two hours. The hanger isn't a cold pitch — it's the follow-up to something they already witnessed. That context does most of the selling.

The message should be direct and neighborly, not corporate. "We just finished a cleanout next door — if you've got anything piling up, we'd love to help. Mention this hanger for $20 off." That's it. No list of services, no company history. Just relevance, timing, and a reason to act.

Response rates on door hangers are modest — typically around 1% in our experience — but in the context of a service business the math is favorable. A hundred hangers might cost $30 and generate one or two calls. One job covers the cost of thousands of hangers. The real multiplier is consistency — five houses per job, every job, across a full year of work compounds into significant neighborhood saturation.

Quick conversations work alongside hangers. If a neighbor is outside while you're working, a thirty-second introduction goes a long way. “Hey, I'm Justin with Grizzly Junk Pros — we're finishing up next door. If we've been in your way at all, I apologize, and here's my card in case you ever need us.” You're not selling. You're being a decent person who happens to be in their neighborhood. Those brief interactions plant seeds. Occasionally they turn into immediate jobs — “actually, we have a shed we've been meaning to clear out.” More often they plant a name that surfaces when the need eventually arrives.

Always respect no-soliciting signs. This approach works because it feels neighborly. The moment it feels pushy it loses all its value.

Yard Signs: Free Billboards on Your Best Customers' Lawns

A yard sign at a completed job is a live endorsement. A homeowner who lets you stake your sign in their lawn is telling every neighbor who drives past that they trusted you with their property. That implicit social proof is worth more than any ad you'll ever write.

The mechanics are simple. Keep a few wire-stake signs in every truck. When a job goes well, ask the customer if you can leave a sign for a few days — most say yes, especially if you frame it as a small favor. Position it where it's visible from the street. Pick it up when you said you would or let them dispose of it on their own.

Dumpster branding does this automatically. Every rented container sitting in a driveway for a week with your name on it is a yard sign that your customer is already paying for. Make the branding bold enough that it's readable from a passing car, not just from the curb.

Over time, if you're running multiple jobs in the same area, you'll start to see a compounding effect. One sign generates a conversation. That conversation generates a referral. That referral is another job in the same neighborhood, another sign, another set of neighbors who see your name. This is how you build dominant local brand presence without an ad spend — you earn it job by job in specific geographies until you're the name everyone on the block knows.

Design the signs to be legible, not creative. Company name large. Phone number large. One short line if you have room — "Same-Day Junk Removal" or "Dumpster Rentals Available." High contrast so it reads from a moving car. That's all it needs to do.

Sticky Notes: Low Cost, Surprisingly Memorable

A sticky note on a door feels like a personal message. It doesn't look like an ad. It triggers curiosity in a way a glossy flyer doesn't — someone finds a handwritten-looking note on their door and wants to know what it says before they even register it as marketing.

The message should feel personal and specific. "Just finished a cleanout at your neighbor's — ask for the neighbor discount if you want to clear anything out. John, 555-1234." The small format forces brevity, which makes it feel direct and genuine rather than like a pitch.

1-800-GOT-JUNK? used sticky notes in the early days when they couldn't afford conventional advertising. It wasn't because it was a brilliant brand strategy — it was cheap and it worked. The psychology behind it is simple: curiosity, personalization, and low perceived pressure. People who would throw away a glossy flyer without reading it will read a sticky note because it doesn't look like an ad.

Practical notes on execution: they don't stick well in wind or rain, so weather matters. Avoid mailboxes — it's illegal in most jurisdictions. Doorframes and gate handles work. Keep the volume reasonable — a few per job, targeted at immediate neighbors, not a mass distribution campaign. They're a complement to door hangers and yard signs, not a replacement.

Swag: Give Things People Actually Keep

Branded merchandise works when you give things people genuinely use. A fridge magnet stays on the refrigerator for years. A tape measure lives in a junk drawer for decades. A quality t-shirt gets worn. Each time the item gets used, your name is in front of that person or anyone in their home.

Research consistently shows that promotional products generate strong retention — the majority of recipients keep them for over a year and remember the brand *(Advertising Specialty Institute)*. The key word is practical. Cheap pens get thrown away. Useful items stay.

For a service business targeting homeowners and contractors, strong options are fridge magnets with your number prominently displayed, tape measures, work gloves, and notepads. For a younger demographic or apartment dwellers, reusable bags work well. For contractors specifically, anything tool-adjacent — a mini multi-tool, a koozie for the job site — lands better than something household-oriented.

Leave a small item with every completed job. It costs a few dollars and keeps your name visible in their home until the next time they need what you do. If even 10% of customers who get a fridge magnet call you again or hand your number to a neighbor, the math is strongly positive.

QR Codes: Bridge the Physical and Digital

People got used to QR codes during the pandemic — restaurant menus, payment apps, and contact sharing made scanning second nature across demographics. For a service business, QR codes turn every physical touchpoint into a direct path to booking, reviewing, or learning more.

On a door hanger: "Scan for $25 off your first job" converts curiosity into a tracked lead. On a yard sign: "Scan to see what we did here" links to a before-and-after photo of the job — social proof delivered in real time to a neighbor who's standing in front of your customer's house. On a dumpster: "Scan to reserve one for your project" captures a prospect at exactly the moment they're thinking about needing one.

For review generation, a QR code on a job completion card — "Loved our work? Scan to leave us a quick review" — makes the ask frictionless and dramatically improves follow-through.

Use dynamic QR codes rather than static ones. Dynamic codes let you change the destination URL without reprinting — if you want to redirect the code on your door hangers from a discount offer to a new promotion, you change it in the app rather than reprinting 5,000 hangers. Most QR code generators offer dynamic codes for a small monthly fee that's easily justified by the flexibility.

Always add context above the code. "Scan for $25 off" outperforms a bare code because people want to know what they're scanning before they point their phone at it.

Adapting to Your Market

The tactics are the same across urban, suburban, and rural markets. The execution adapts.

In dense urban environments, door hangers at individual homes work less often because you're dealing with apartment buildings and limited residential access. Shift the emphasis to truck visibility — a well-branded vehicle moving through high-traffic urban streets generates significant impressions. Sticky notes and flyers in building lobbies, laundromats, and community boards work where individual door approaches don't. QR codes resonate more strongly with urban demographics who are comfortable scanning. Building manager relationships are worth pursuing — a single property manager responsible for a hundred units is a better prospect than a hundred individual homeowners.

In suburban markets, you're in the sweet spot. Single-family homes, accessible driveways, neighbors who talk to each other, subdivision entrances where a parked truck gets maximum visibility. Yard signs compound neighborhood by neighborhood. Door hangers work well. Quick chats are natural. The suburban market is where consistent guerrilla tactics build the strongest compounding brand presence over time.

In rural markets, volume is lower but community density is higher. One job done well for someone well-connected in a small town generates referrals that a dozen suburban jobs might not. Community involvement — sponsoring local events, showing up at the county fair, supporting a local cleanup effort — builds the kind of trust that drives word-of-mouth in tight-knit communities. Yard signs stand out more on a rural road with fewer competing signs. The hardware store bulletin board and the diner community board reach people who don't necessarily find you on Google.

Regardless of market: urban customers tend to value efficiency and ease, suburban customers respond to community presence and social proof, rural customers respond to authenticity and relationship. The same tactic deployed with those different emphases lands differently in each context.

Build It Into the Routine

The reason most businesses don't execute on guerrilla marketing consistently isn't skepticism — it's that it feels like extra work on top of an already full day. The fix is making it part of the job completion process rather than a separate activity.

Keep a stack of door hangers and a yard sign in every truck. At job completion, the crew hits five doors and asks the customer about leaving a sign before they pull out. The QR code on the job receipt card goes to the customer as they pay. The swag item comes out with the final walkthrough. The sticky notes go up while the last load is being secured. None of this adds more than ten minutes to the job. Multiplied across every job for a full year, it adds up to thousands of neighborhood-level brand impressions that compound into calls you'd never get from digital alone.

Track what you can. Ask every new caller how they found you. When customers mention the truck or the door hanger or the yard sign, you get confirmation that the effort is working and which specific tactics are producing results in your specific market.

The businesses that seem to be everywhere in their local market didn't necessarily get there by running bigger ads. They got there by showing up consistently in small ways across every job, every neighborhood, and every day — until their name felt familiar to everyone in a five-mile radius.

> *The best time to market your business is while you're already doing the work.*

Go Deeper

For more on this and everything else in the book, the Haulers' Edge Newsletter goes deeper every week. Scan the code or subscribe free at HaulingHubb.com.

CH. 16 STOP WINGING IT WHEN THEY CALL

"Every sale has five basic obstacles: no need, no money, no hurry, no desire, no trust." — Zig Ziglar

The phone rings. Someone needs junk removed or a dumpster dropped. They're ready to book. And the person who answers — whether it's you, your admin, or a crew member who happened to pick up — wings it.

No script. No process. No consistent way of asking the right questions, quoting with confidence, or closing the job. Just a conversation that goes however it goes, ends however it ends, and produces whatever result luck and personality happen to deliver that day.

This is how most service businesses handle their inbound calls. And it's costing them more than they realize.

Your marketing dollars, your Google Ads, your yard signs, your five-around door hangers — all of that effort exists to make the phone ring. The call is the moment all of it either pays off or doesn't. A strong phone process converts that lead into a booked job. A weak one sends them to your competitor, who answered more confidently and made it easier to say yes.

Why the Phone Still Matters More Than Most Operators Think

In a world of online booking and automated everything, it's tempting to assume the phone is becoming less important. For local service businesses, that assumption is wrong — and the data keeps proving it.

According to BrightLocal, 60% of consumers prefer to contact local service businesses by phone after finding them online. For home services specifically — anything involving access to someone's property — the conversion rate on phone inquiries is significantly higher than web form submissions. Industry data consistently shows phone inquiries convert at roughly two to three times the rate of web form submissions for service businesses, because a caller who's ready to talk is further along in their decision than one who's still browsing. When someone calls, they've already made a significant mental commitment. They're not browsing. They're ready to hire someone. Your job is to make sure that someone is you.

The phone call is also where trust gets established fastest. A website can look professional. A Google profile can have great reviews. But a real conversation with a competent, confident human who quickly understands their problem and gives them a clear path forward converts at a rate no web form can match. A bad call experience — fumbling for information, vague pricing, long pauses, no clear next step — destroys trust that your marketing spent real money building.

Every call that doesn't book is revenue you paid to generate and then failed to close. Tracking your call-to-booking conversion rate and improving it by even 10% produces more revenue than most marketing investments you'll make.

The Winning Call Framework

A phone call that reliably converts follows a consistent structure. Not a robotic script — a natural process with a clear shape. The Winning Call Framework has five stages. Here's what each one looks like.

Stage 1: Answer fast and set the tone immediately.

The first three seconds of a call establish whether the prospect feels like they reached a professional operation or a chaotic one. Pick up within three rings. Answer with your business name and your name — "Grizzly Junk Pros, this is Justin, how can I help you?" That one line tells them they reached the right place, there's a real person on the line, and this business operates professionally.

Voicemail is a conversion killer. If you can't answer, your voicemail message matters — "You've reached Grizzly Junk Pros. We're on a job right now but will call you back within the hour. Text us photos of what needs to go for a faster quote at [number]." That message shows responsiveness, sets an expectation, and gives them something to do that keeps them engaged rather than calling your competitor.

If you're using an after-hours AI voice agent or automated response system — covered in Chapter 32 — make sure it captures the lead and triggers an immediate callback rather than just taking a message. Speed of follow-up is a direct determinant of conversion rate.

Stage 2: Understand the job before you quote it.

The instinct of most operators is to get to a number as fast as possible. Resist that. Before you quote, you need to understand what you're quoting. Rushing to a number without understanding the job leads to either underquoting — which costs you money — or overquoting — which loses the job.

Ask three to four questions that give you what you need. What are you looking to get rid of? Roughly how much is there — a few items, half a truckload, a full truck? Where is it located — in the house, garage, outside? Is there anything that might need special handling — appliances, electronics, tires, hazmat?

These questions do two things simultaneously. They give you the information to quote accurately. And they show the customer that you're thorough and professional — that you actually understand their job rather than just throwing a number at them. Most of your competitors skip this step. The operator who takes sixty seconds to understand the job before quoting it sounds markedly more competent than the one who immediately says "probably around $300 or so."

Stage 3: Quote with confidence and clarity.

After you understand the job, give a clear number or range. Not "it depends" — an actual figure. "Based on what you're describing, that's probably going to run between $350 and $450 for us, with the final number confirmed when we see it. That includes all the labor, the truck, and disposal — nothing extra."

Three things in that quote that matter. A specific range, not a vague "it depends." What's included, stated explicitly. And a brief explanation of why the final number gets confirmed on-site — which sets the expectation correctly rather than feeling like a bait and switch later.

If the job is clearly priced by your standard rate — a single-item pickup, a standard dumpster rental, a flat-rate service — give the flat number confidently. Hedging on a price you know makes you sound unsure of your own rates. State it clearly, pause, and let them respond.

Don't negotiate against yourself. Quote the number, stop talking, and wait. Most operators fill the silence by immediately explaining why the price is what it is, which signals that they're not sure the price will hold. Say the number. Stop. Let them react before you say anything else.

Stage 4: Handle the objection without flinching.

The most common objections on a service business call are: that's more than I expected, someone else quoted me less, and can you do better on the price.

For "that's more than I expected": don't panic, don't apologize, and don't immediately discount. "I understand — what were you expecting?" That question does two things. It gives you information about where the gap is. And it buys you a moment to respond specifically rather than reactively. Once you know what they expected, you can address it directly — either by explaining what's included in your price that may not be in their expectation, or by finding a scaled-down option that fits their budget without destroying your margin.

For "someone else quoted me less": "That's possible — there are cheaper options out there. Just make sure the quote includes disposal fees and that they're fully insured, because some lower quotes don't cover those. Our price is all-in with no surprises." You've planted a legitimate question about the competitor's quote without saying anything negative about them directly. You've also reinforced what they get with you — no surprises, full coverage.

For "can you do better": "Our pricing is based on the actual cost of doing the job right — labor, disposal, insurance. We don't have a lot of room to move on that. What I can do is offer something specific — schedule it this week, prioritize your preferred date, or throw in a small add-on at no charge." You're not budging on price. You're offering something that has real value to them without eroding your margin.

The underlying principle: confidence holds price better than any sales tactic. An operator who knows their numbers, knows what the job is worth, and communicates that clearly without apology will close more jobs at better margins than one who discounts reflexively the moment there's any pushback.

Stage 5: Close and confirm.

Once the customer is ready, make the next step completely clear and handle it immediately. "Great — let me get you on the schedule right now. What day works best for you?" Don't end the call with "I'll send you a quote" or "give us a call back when you're ready." Close on the call. Every step between the end of the conversation and the booking is an opportunity for them to call someone else.

Capture their name, number, address, and a brief description of the job. Confirm the date and time window. If you take a deposit for large jobs, collect it before the call ends — "I can take a credit card right now to hold that spot." Repeat the key details back to them at the end: "So we've got you down for Tuesday between 8 and 10am at [address], full garage cleanout, roughly half to full truck. Does that all sound right?" That confirmation loop catches misunderstandings before they become problems on job day.

End with a clean close: "We'll send you a reminder the day before. Looking forward to it." Brief, professional, done.

Beyond the call itself, two habits consistently improve conversion before and after the conversation.

Ask More Than You Talk

The best salespeople in any industry share one trait that has nothing to do with charisma or persuasion. They ask more questions than anyone else in the room and they listen to the answers with genuine attention. In a service business that skill is the difference between booking a job and building a customer.

Most intake calls follow the same pattern. Customer describes what they need. Operator quotes a price. Customer says they'll think about it or books on the spot. That transaction captures the obvious job and misses everything around it — the real pain point driving the call, the adjacent need the customer hasn't mentioned yet, the specific outcome they care about that has nothing to do with the service itself.

The customer calling about a garage cleanout isn't always

calling because they have junk. Sometimes they're calling because their mother-in-law is moving in next month and the garage needs to become a bedroom. Sometimes they're calling because they're listing the house and the realtor told them the clutter is costing them ten thousand dollars on the sale price. Sometimes they're calling because they've been staring at that garage for three years and they're finally doing it and they just need someone to make it easy. The job is the same in all three cases. What the customer cares about is completely different — and if you know which one you're dealing with, you can speak directly to it.

Ask questions that surface the real motivation. "What's prompting this now — is there a timeline you're working toward?" "Is there anything specific about how this goes that matters most to you?" "Has anything like this been done before or is this the first time?" These aren't interrogation questions. They're the questions a good contractor asks before they start work because they actually want to do the job right. Customers feel the difference immediately.

Take notes during the call. Not just the job details — the things the customer said that revealed what they actually care about. If they mentioned the realtor, your crew shows up knowing this job is about the house sale and the customer experience they deliver reflects that. If they mentioned the mother-in-law, the crew knows there's a timeline and sensitivity attached. The customer who feels like the company remembered what they said — because the notes are in the job record and the crew was briefed — becomes the customer who leaves the five-star review that mentions how the team went above and beyond.

Upsells surface naturally in this kind of conversation. The customer who mentions they're listing the house probably also needs the basement cleared. The customer who's finally tackling the garage after three years probably has other spaces they've been avoiding. You don't have to manufacture the upsell opportunity — you just have to listen long enough for the customer to tell you where it is. "While we're there, would it help to take a look at the basement as well?" is a natural offer when the conversation has already told you the basement exists and matters.

Going above and beyond doesn't always require doing more work. Often it just requires showing up already knowing what the customer cares about and making decisions on the job accordingly. Leave the garage broom-swept because you knew they were listing the house. Move efficiently because you knew the mother-in-law arrives Friday. Those are not extraordinary gestures — they're the natural result of listening on the intake call and acting on what you heard.

The customer who feels genuinely heard before the job starts is the customer who writes the review that says the team felt like they actually cared. Because they did. And they showed it by asking the right questions and remembering the answers.

Send the Answer Before They Ask the Question

The best sales conversation you'll ever have is one where the customer already knows the answers to their biggest questions before you say a word.

Most operators wait for the call to do all the selling. The prospect calls cold, you qualify them, you quote, you handle objections, you try to close. Every step requires you to build trust from scratch in real time while the customer is simultaneously evaluating whether to trust you at all.

There's a better approach. The night before every estimate call or job confirmation send the customer something that answers their most common questions before they can ask them. A short video — two minutes, filmed on your phone — that walks them through exactly what to expect. How the process works. What the crew will do when they arrive. What affects the final price. What you won't take and why. What happens at the end of the job.

That video does three things before you ever pick up the phone. It establishes you as a professional operation that communicates clearly. It answers the objections that would have slowed down the call. And it creates a customer who arrives at the conversation already sold on the experience rather than still deciding whether to trust you.

The conversion rate on calls preceded by that kind of pre-call content is consistently higher than cold calls — not because the video is persuasive, but because it shortens the education curve and builds trust before anyone steps foot on the property. The customer who watches it arrives at the conversation already understanding your process, already knowing what to expect, and already more comfortable with you as a person. The tire kickers often eliminate themselves before you spend fifteen minutes on a call that was never going to close. What's left is a shorter, warmer conversation with someone who's already most of the way there.

Set this up once. Film a two-minute walkthrough of your process. Send it via text the evening before every estimate. Track whether your close rate changes over the following ninety days. It will.

Build a Script That Doesn't Sound Like a Script

A script isn't a word-for-word recitation. It's a framework — the key questions to ask, the key information to capture, the key phrases that handle common objections, and the clear path to close. Once someone knows it well enough, it sounds like a natural conversation. Until then, having it written down prevents the fumbling and forgetting that kills conversion.

Write out your call script in plain conversational language. Your opening. The four qualifying questions. Your standard quote language for the two or three most common job types. Your three objection responses. Your close and confirmation sequence. Print it out and keep it near every phone. Train anyone who answers calls on it until it's natural.

The script should also capture the information you need for booking. Build a simple intake form — physical or digital — that the person on the phone fills out during the call. Name, number, address, job description, preferred dates, how they found you. That form feeds your CRM, creates the job record, and ensures nothing gets lost between the call and the job.

Here's a small technique that consistently works for getting customer contact information without feeling like a sales pitch. Most people hesitate to give out their email because they've been burned before — they hand it over and immediately start getting promotional emails they never asked for. So don't ask for it the way everyone else does.

Instead, take a job away from them.

At the end of the call, before you hang up, say this: "Would you like me to send everything we just discussed to you in an email so you have it in writing and don't have to write anything down?" Almost everyone says yes. And why wouldn't they — you just offered to do work for them. You're removing the task of memorizing the price, the appointment time, the scope of the job, and any other details from the conversation. You're making their life easier.

The result is you now have their email address, they gave it willingly, and they're actually glad you asked. From there your follow-up sequence runs automatically — confirmation, reminder, post-job review request, seasonal check-in. The entire customer relationship is now connected through a channel they handed you because you framed it as a service rather than a data grab.

It also signals something important to the customer before the job even starts. You're organized. You follow through. You communicate. That impression is set before your truck pulls into the driveway — and it's set because of one sentence at the end of a phone call.

Who Answers Your Phones Matters

In most small service businesses, the phone gets answered by whoever is available — the owner, an admin, a crew member, sometimes nobody. That inconsistency produces inconsistent results.

If you have an admin handling calls, they need the same training and the same script as you would use. They need to know your service area, your pricing ranges, your common job types, and how to handle the four or five objections they'll encounter daily. A well-trained admin consistently following a strong process will out-convert a distracted owner improvising on every call.

Whoever answers phones should not be multitasking in a way that's audible. Wind noise, equipment noise, driving — all of it signals to the caller that they don't have your full attention. If you can't give the call full attention right now, let it go to voicemail with a message that sets a fast callback expectation, and call back within the hour.

Track who answers calls and what the conversion rate is by person. If one team member converts 70% of calls and another converts 30%, the difference is almost always process and confidence, both of which are trainable. Record calls if your jurisdiction allows — with customer disclosure where required — and use them for training. Hearing their own calls is the fastest way to help someone improve.

The person who answers your phone is the first human being your customer encounters. How they sound, how they engage, and whether they feel like a real person who cares about the caller's situation determines whether the conversation builds trust or just processes a transaction.

The Missed Call Is an Emergency

Every missed call in a service business is a lead that cost money to generate, got to the moment of conversion, and disappeared.

The standard service business behavior when a call is missed: the caller gets voicemail, maybe leaves a message, and waits. Meanwhile they open Google again and call the next result. By the time you call back two hours later, they've already booked someone else.

Treat every missed call as a time-sensitive event. Call back within fifteen minutes. If you're on a job and can't call, have your CRM trigger an automatic text immediately — "Hey, this is Justin at Grizzly Junk Pros — sorry I missed you. I'll call you back within the hour. In the meantime, text me what you need and I'll get you a quick estimate." That text does three things. It shows the caller they weren't ignored. It starts a conversation thread. And it often captures enough information to quote by text, which converts a meaningful percentage of leads who don't want to wait for a call.

According to research by InsideSales.com, businesses that respond to leads within five minutes are 100 times more likely to contact and qualify the prospect than those that respond after thirty minutes *(InsideSales.com, Harvard Business Review, 2011)*. That gap is enormous. Speed of response is one of the single highest-leverage things you can improve in your conversion process.

AI Voice Agents: Never Miss Another Lead

There's a practical limit to how many calls a small service business can answer in real time. You're on a job. Your admin is on another call. It's 9pm. The phone rings and nobody picks up.

Historically that meant voicemail — and voicemail meant a meaningful percentage of those callers moved on before you called back. Now there's a better option.

An AI voice agent answers every call your team can't get to — overflow during business hours and everything after hours — and handles the interaction in a way that keeps the lead alive until a real person follows up. The one principle that makes or breaks the setup: the agent must identify itself as AI immediately and clearly. That honesty builds trust rather than eroding it — callers told upfront almost always stay on the line, because the alternative is voicemail or hanging up and calling your competitor.

For the full build guide — how to train the agent on your services and pricing, how to script it, and how to connect it to your pipeline so leads flow in automatically — see Chapter 32.

Track Your Conversion Rate

Most service business owners have no idea what percentage of their inbound calls convert to booked jobs. They know roughly how busy they are, but they can't tell you whether their phone process is converting 40% of calls or 70%.

That number matters because it tells you how much revenue you're leaving on the table. If you're converting 40% of inbound calls and you improve that to 60%, you've grown your revenue by 50% from the same volume of calls — without spending an additional dollar on advertising.

Before you build anything else, run the math on your current close rate. If you're answering 100 calls a month and closing 60% at an average job value of $400, that's $24,000 in booked revenue. Moving to 70% — one extra booking for every ten calls — produces $28,000. That's $48,000 a year from a process improvement that costs nothing. At 75% it's $30,000 monthly, $6,000 more than where you started. A scripted, trained, measured call process is not about sounding less natural. It's about converting more of the leads you already paid to generate.

Track it simply. Log every inbound call — in your CRM or even a spreadsheet. Note whether it converted to a booking. Review the number weekly. If your conversion rate drops, investigate — did someone new start answering calls, did a competitor undercut on price, is there a specific objection coming up repeatedly? If it climbs, understand why and reinforce whatever changed.

Call tracking tools like CallRail — mentioned in Chapter 14 — give you the data to do this analysis per keyword, per ad, and per person answering the phone. At a few dollars per day, the insight they provide is worth far more than their cost.

When the Call Doesn't Close

Not every call will book. Some prospects are genuinely just gathering information. Some have a budget that doesn't match your pricing. Some are comparison shopping and will call three companies before deciding. That's the reality of a service business.

For the calls that don't close immediately, capture the information and follow up. If they said they'd think about it, send a text within two hours: "Great talking to you earlier. Let me know if you have any questions — I held a spot tentatively for Tuesday if you want to go ahead." That follow-up converts a meaningful percentage of maybes.

For the calls that end with "you're too expensive," don't chase them with a discount. Your price is built on real costs. Discounting to win a price-shopper sets a precedent and attracts customers who will price-shop you again every time. Let them go. The customer who books you at your full rate and calls you back twice a year is worth ten times the customer who negotiated you down and will do it again at every interaction.

The call process isn't about winning every lead. It's about converting the right leads efficiently and consistently — so the marketing that generates them produces the maximum possible return.

Record your next five inbound calls. Listen to them with the Winning Call Framework from this chapter in front of you. You'll hear exactly where the process breaks down — and you'll know exactly what to fix.

Every ring is an opportunity. Answer it like one.

Close rate by source and no-show rate — the two metrics that tell you whether your phone process is working — are in your weekly scorecard in Appendix A.

Go Deeper

For more on this and everything else in the book, the Haulers' Edge Newsletter goes deeper every week. Scan the code or subscribe free at HaulingHubb.com.

CH. 17 THE MONEY IS IN THE FOLLOW-UP

"Patience, persistence, and perspiration make an unbeatable combination for success." — *Napoleon Hill*

You spent money to generate the lead. Your Google Ad got clicked. Your yard sign got noticed. Your crew did good work and the neighbor called. Someone asked for a quote, seemed genuinely interested, and then went quiet.

Most operators send one follow-up — maybe — and then move on. The lead dies in their inbox and they assume the customer chose someone else or decided not to do the job. Sometimes that's true. More often, life just got in the way. They got busy. The estimate slipped down the email thread. They meant to call back and forgot.

The business that follows up consistently gets those jobs. The one that doesn't, doesn't.

That gap between the operators who follow up and the ones who don't is where a significant amount of revenue is hiding.

Why Most Leads Don't Close on First Contact

Research consistently shows that fewer than 5% of sales happen on the very first contact *(Marketing Donut, widely cited)*. The majority require multiple follow-ups — yet nearly half of all businesses make only one attempt and stop *(Scripted/Salesforce research, widely cited)*. That means almost half your competitors are leaving money on the table every single day by quitting after the first no-response.

For service businesses specifically, the dynamic is predictable. A homeowner gets your quote, compares it mentally against a vague idea of what they expected to spend, gets distracted by the day, forgets about it, and moves on. They haven't chosen your competitor — they've just gone back to their normal life. A well-timed, well-worded follow-up interrupts that drift and brings the decision back to the surface at a moment when they might actually act.

The math on this is worth sitting with. If your average job is worth $400 and you're generating 20 estimates a month with a 30% close rate, you're booking 6 jobs. Improve your close rate to 45% through consistent follow-up and you're booking 9 jobs — an extra $1,200 a month from the same marketing spend, the same number of estimates, just a better follow-up process.

Speed matters as much as persistence. Research shows that responding to a new inquiry within five minutes makes you 100 times more likely to contact and qualify the prospect than waiting thirty minutes *(InsideSales.com, Harvard Business Review, 2011)*. Same thing applies to estimates — a follow-up sent the same day or the next day catches people while your quote is still in their mental queue. Wait a week and you're starting from scratch.

Read the Customer Before You Run the Sequence

Not every lead has the same urgency and your follow-up strategy should reflect that. A customer who calls asking if you can be there tomorrow is in a completely different place than one who mentions they're planning a renovation next spring. Treating them identically — same cadence, same timing, same messaging — misses the point of follow-up entirely.

When a lead comes in, gauge their intent and urgency before the sequence starts. A few signals tell you almost everything you need to know. Did they ask about availability this week? Did they mention a specific deadline — a move, a closing date, a contractor starting Monday? Are they asking detailed questions about pricing and process rather than just browsing? High urgency customers need tight follow-up — same day, next morning, a direct offer to hold a slot. Every hour you wait on a high-urgency lead is an hour they're calling the next company on the list.

Low urgency customers need a different approach entirely. Someone planning a project three months out doesn't need you calling them twice in forty-eight hours — that follow-up strategy will annoy them out of your pipeline before the job is ever relevant. They need a longer, more patient cadence. A check-in at two weeks. A seasonal reminder when the timing gets closer. A useful piece of information that keeps you present without being pushy. The goal with a long-runway lead isn't to close them today — it's to be the company they think of when the project becomes real.

Service Hubb AI and some other CRMs lets you tag leads by urgency at intake and route them into different follow-up sequences automatically. High urgency gets the tight Three-Touch System from earlier in this chapter. Long-runway leads go into a slower drip that stays in contact over weeks or months without wearing out the relationship. One system, two strategies, applied based on what the customer actually told you about their timeline.

The Three-Touch System

You don't need a complicated system. You need a consistent one. The Three-Touch System — three contacts over thirty days — handles the vast majority of convertible leads.

Day one or two — the confirmation touch. Send a short message within 24 hours of the estimate. Not a sales pitch — a confirmation. “Hey [Name], just wanted to make sure the estimate came through okay and that everything looked clear.

Happy to answer any questions." That's it. You're showing responsiveness, you're giving them an easy on-ramp to ask questions they might have been too busy to ask before, and you're re-establishing that you're a real person who pays attention.

Day five to seven — the value touch. If you haven't heard back, reach out again with something more than a check-in. Reference the specific job. Mention something relevant — your availability is opening up for that week, you just finished a similar job nearby, you have a question about one detail of their project. "Hey [Name] — following up on the cleanout quote. We wrapped up a similar garage job in your area last week and it took us about three hours. Wanted to make sure you had everything you needed before making a decision." You're not asking for a yes. You're showing you remember their project and you're still available.

Day thirty — the door-open touch. One final message that takes the pressure completely off. "Hey [Name], just wanted to check in one last time on the estimate we sent over. If the timing hasn't been right, no problem at all — we'll be here when you're ready. Just keep our number handy." That message does something most businesses never do — it gives the prospect explicit permission to not be ready, which paradoxically makes them more likely to respond. And even if they don't book now, you've left a positive last impression that makes them think of you first when the timing does become right.

Three touches. Thirty days. Consistent across every estimate you send.

Channel Matters as Much as Timing

Different customers respond to different channels. An email that sits unread for a week might get a text response in four minutes. Average email open rates for service businesses run around 20–25% *(Mailchimp industry benchmarks)*. SMS open rates run 95–98%. That gap is the whole argument for text as your primary follow-up channel. A phone call that goes to voicemail might prompt a text reply an hour later. The operators who follow up across multiple channels consistently outperform those who stick to one.

Email works best for the estimate itself and for messages that include detail — photos, a breakdown of what's included, links to your reviews. The downside is inbox competition. An email can sit buried under twenty others and never get opened.

Text is your highest-open-rate channel. Studies consistently put SMS open rates at 95–98%, with most messages read within minutes *(SimpleTexting, 2023)*. For a follow-up message that's two or three sentences, text is almost always the right choice. Keep it conversational — not corporate. “Hey [Name], this is Justin from Grizzly Junk Pros — just checking in on that cleanout quote. Any questions I can answer?” reads like a real person. A formal paragraph reads like a form letter.

Phone calls are the most personal option and work particularly well for the second or third touch, especially on larger jobs where trust is a bigger factor. Even a voicemail matters — hearing a real voice behind the name on the estimate builds trust in a way text can't. Keep voicemails short and specific: “Hi [Name], this is Justin with Grizzly Junk Pros — following up on the garage cleanout estimate from Tuesday. We've got openings next week if you'd like to move forward. Call or text me back at [number].”

The most effective sequence for most service businesses: email the estimate, text the day-one confirmation, text or call on day five to seven, text the day-thirty close. Adjust based on how the customer communicated with you initially — if they texted you to request the estimate, stay in text. If they called, call back.

Service Hubb AI handles this sequencing automatically. When

an estimate goes out, the follow-up sequence triggers — the right channel, the right timing, the right message — without you manually tracking who needs a follow-up and when. The lead doesn't fall through the cracks because the system doesn't forget.

Make Every Follow-Up Feel Personal

The fastest way to get ignored is sending a generic "just checking in" message three times in a row. Customers can feel the difference between a template and a message written for them. It takes thirty seconds to add one specific detail that transforms a generic follow-up into something that feels personal.

Reference the actual job. Use their name. Mention something they told you — the graduation party they wanted the yard cleared for, the contractor who's waiting on the debris to be removed, the basement that's been cluttered since the kids moved out. That detail shows you were paying attention and that you see their situation specifically, not just a generic job opportunity.

"Hi Sarah — following up on the estimate for the backyard. You mentioned you wanted it done before your daughter's graduation party in June. We've got a crew available the first two weekends of May if you want to get that locked in before our calendar fills up." That message gets opened and read. "Hi, just following up on the estimate we sent" gets deleted.

Adding value works the same way. If you have a relevant before-and-after photo from a similar job, send it. If there's a question they might be sitting on that you can answer proactively, answer it without waiting to be asked. "By the way — the piano you mentioned, we can handle that, we just add a two-person surcharge for anything over 200 pounds. Wanted to clarify that before you compared quotes." That kind of detail demonstrates competence and removes a potential obstacle without them having to ask.

Not Every Email Should Be About the Job

The email nurture sequence that runs entirely on promotional content — reminders to book, seasonal offers, referral requests — trains your list to tune you out. When every message is asking for something, readers stop opening them. When they do open them and find another pitch, they unsubscribe.

The sequences that actually build relationships include content that has nothing to do with selling. Genuinely useful information that makes the customer's life slightly better — delivered under your name, in your voice, on a consistent schedule. That's what keeps you in their inbox without annoying them out of it.

For a home service business that content might look like a seasonal home maintenance checklist relevant to your area. Local events worth knowing about. A tip about how to prepare for a renovation that has nothing to do with whether they hire you. A heads-up about a regulation or program in their area that affects homeowners. Information about donation centers or recycling options in your market. None of it is a pitch. All of it is useful. And every time they open it your name gets another impression with zero resistance.

The psychology is straightforward. People don't mind hearing from businesses that consistently provide value. They deeply resent hearing from businesses that only reach out when they want something. By mixing genuinely helpful content into your nurture sequence — one value email for every promotional email at minimum — you train your list to open your messages because opening them is usually worth it. When you do send a promotional message it lands in an inbox that's already receptive rather than one that's been conditioned to ignore you.

The customers who are six months away from needing your service will remember the company that sent them something useful every few weeks. They won't remember the one that went quiet after the estimate and then reappeared with a discount code. Stay present. Stay useful. Stay in their inbox with something worth reading — and when the job becomes real, you'll be the first call they make.

Urgency That's Real, Not Manufactured

A gentle sense of urgency helps people stop procrastinating. Most customers who haven't responded to your estimate aren't saying no — they're saying not yet. A legitimate reason to act now moves them off the fence.

Real urgency comes from real constraints. Your schedule filling up for the month. A price increase going into effect next week because disposal fees went up. A weather window closing for an outdoor job. If those things are true, say so — it creates a reason to act that doesn't feel like pressure. "We've got two openings left this month before we're booked solid for summer — want me to hold one for you?" is legitimate if it's accurate.

What doesn't work is invented urgency. Fake deadlines, fake scarcity, fake price increases — customers sense these and they destroy the trust you've been building with every professional interaction up to that point. One manufactured pressure moment can undo three touchpoints of goodwill.

The goal is never to pressure someone into a decision they're not ready to make. It's to give someone who's already leaning toward yes a reason to move today instead of next month.

Know When to Stop

After three to four follow-ups with no response, send a final exit message and move on. "Just wanted to close the loop — totally understand if the timing hasn't been right. Our estimate stands whenever you're ready, and we'd love to help when the time comes." Then stop reaching out unless they initiate.

This serves two purposes. It gives you closure on the lead so it doesn't sit indefinitely in your pipeline. And it sometimes triggers a response from customers who felt slightly pressured and will respond positively to being let off the hook — "actually, let's go ahead and book it."

The customers who don't respond after four thoughtful follow-ups across thirty days either aren't ready, aren't a fit, or have already hired someone else. Respecting their silence and ending cleanly is the professional move. Some of them will come back six months later when the timing is right, remembering that you followed up persistently but never made them feel harassed.

Build the System So It Runs Without You

The reason most operators don't follow up consistently isn't because they don't know they should. It's because they're running a service business and following up on twenty estimates while managing jobs, crews, and customers requires a system or it doesn't happen.

Service Hubb AI handles the follow-up sequence the same way it handles lead capture from the website and AI voice answering — automatically, based on rules you set once. When an estimate is sent, the sequence starts. Day two confirmation goes out. If no response, day seven value touch fires. If still no response, day thirty close goes out. Every message is templated in your voice but personalizable with job-specific details where you've added them. The ones that get responses automatically pause the sequence and notify you so you can take over the conversation.

For operators who aren't yet using an automated system, a

simple spreadsheet works — name, estimate date, last follow-up date, next follow-up date, channel used, status. Ten minutes at the start of every day reviewing that sheet catches every lead that needs a touch. It's not elegant but it works, and it's infinitely better than relying on memory.

The key is that following up cannot be something you do when you remember. It has to be a system that runs regardless of how busy the day gets. The $400 job that closes because you sent a day-seven text you almost skipped is revenue you would have left behind if the system hadn't reminded you.

The Revenue Math

If you're running a typical service business and answering calls consistently but doing no structured follow-up on estimates, industry data suggests you're converting roughly 30–40% of your inbound leads *(general service industry benchmarks)*. Add the Three-Touch System and that rate typically improves by 15–25 percentage points. Three touches is the minimum — the 80% of sales that require more contact are why the sequence doesn't stop at one or two.

On 20 estimates a month at an average job value of $400, the difference between a 35% close rate and a 50% close rate is three extra jobs — $1,200 a month, $14,400 a year. From the same marketing spend. From the same estimate volume. Just from following up instead of waiting.

That's not a marginal improvement. It's a meaningful revenue line that most businesses are leaving uncollected because nobody built the follow-up system.

The fortune is genuinely in the follow-up. Not as a motivational phrase — as a measurable fact about where money is sitting in your business right now, waiting for someone to go get it.

One prerequisite worth flagging before you leave this chapter: your overall close rate is useful, but your close rate by source is where the real insight lives. Your Google Ads leads and your referral leads close at very different rates — and if you don't know which is which, you're making budget decisions blind. That calculation requires every lead to be tagged at entry in your CRM with its source. Build that habit now, before you have enough data to analyze, so that when you run the monthly review in Appendix A the numbers are actually there.

> *Send the text. Make the call. The job that was going quiet might be one message away from a yes.*

Go Deeper

For more on this and everything else in the book, the Haulers' Edge Newsletter goes deeper every week. Scan the code or subscribe free at HaulingHubb.com.

CH. 18 HOOKING CLIENTS ON REPEAT

"If people like you, they'll listen to you, but if they trust you, they'll do business with you." — Zig Ziglar

The hardest customer to win is the first one. You paid to generate the lead, earned the trust, showed up and did the work. That acquisition cost is real — in ad spend, in time, in the effort it took to convert an estimate into a booked job.

The second job from that same customer costs you almost nothing. They already know you. They already trust you. When they need the service again, or when their contractor friend asks who they use, you're the first name that comes to mind — if you've done the right things to stay there.

Most service businesses don't. They complete the job, move on to the next lead, and effectively start their customer acquisition process over from zero every single day. The operators who build real businesses figure out how to make the work they've already done keep paying.

Chapter 11 built the system that keeps customers from going cold — the follow-up sequences, reactivation campaigns, and referral asks that turn completed jobs into ongoing relationships. This chapter goes a level further. It's about engineering the structure that makes repeat business automatic — packages, contracts, and commercial relationships designed so the customer never needs to make a buying decision again. If Chapter 11 is retention, this chapter is architecture.

The Customers Worth More Than They Look

Not all customers are created equal and the math makes this clear.

A homeowner who calls for a one-time garage cleanout is worth $400. If they call you again in two years for a basement cleanout, refer a neighbor who books two jobs, and mention you to their real estate agent — that original $400 customer is now worth several thousand dollars in lifetime value without you spending a single dollar to acquire any of those follow-on jobs.

A contractor who gives you five dumpster rentals a month at $400 each is generating $2,000 a month from a single relationship. No ads, no estimates, no follow-up sequences. One phone call a week. Compared to the cost of generating twenty individual residential leads to produce the same revenue — the difference is enormous.

The retention math we covered in Chapter 11 compounds differently when the customer has ongoing needs rather than occasional ones.

The clients with the most predictable repeat needs in a service business tend to fall into the same categories across most trades. Contractors who generate debris on every project. Property managers who turn over units on a regular cycle. Real estate investors who buy, renovate, and flip properties continuously. Landlords managing multiple units. Restoration companies who need debris removed after every job. These aren't customers who call once — they're customers who call constantly, provided you make the relationship easy enough that switching to someone else isn't worth the friction.

Building your business around a core of these high-frequency commercial clients while maintaining a healthy residential base creates a fundamentally different revenue picture than chasing individual homeowners one by one. One loyal contractor giving you consistent work is worth more than a hundred one-off residential jobs in acquisition cost alone.

How to Win and Keep Commercial Clients

The first step is actually pursuing them — which most operators don't do systematically. They wait for commercial clients to find them the same way residential customers do, through Google or a referral. But commercial clients are reachable directly and they respond to direct outreach better than most operators expect.

Introduce yourself. Show up at a contractor association meeting or a local landlord group. Connect with property managers on LinkedIn. Send a short email to renovation companies in your service area — not a sales pitch, a simple introduction. "I run a junk removal and dumpster rental company in [city]. We specialize in working with contractors and property managers who need fast, reliable service. If you ever need someone you can count on, I'd love to be on your list." One sentence about who you are, one sentence about who you serve, one sentence about what you're offering. That's it.

The conversion from initial outreach to first job often takes time — they're not going to switch from their current vendor the day you call. But when their current vendor lets them down, which happens, you're the person they remember reaching out. Planting the seed consistently across your market builds a pipeline of commercial relationships that converts steadily over months.

Once you have a commercial client, keeping them is mostly about one thing: reliability. Show up when you say you will. Answer when they call. Solve problems without drama. The bar for loyalty in commercial relationships is not high — most operators lose commercial clients not to competitors who are dramatically better but to competitors who are slightly more consistent. If you're the one who always picks up and always delivers, switching becomes a risk they don't want to take.

Go beyond the minimum on the first few jobs with a new commercial client. Drop the dumpster an hour early. Send a confirmation text the day before. Sweep up without being asked. These small overdeliveries in the first two or three interactions establish a baseline expectation that you operate at a higher level than whoever they used before. That impression is difficult to dislodge once it's formed.

Over time, make yourself structurally useful to their operation. Does the contractor need a dumpster the day a project starts? Be the company that can guarantee that. Does the property manager need guaranteed next-day response on unit cleanouts? Build that into how you serve them. The more deeply your service fits into how they operate, the higher the switching cost for them — and that's what creates real stickiness.

Create Packages That Make the Decision Easy

One of the simplest ways to drive repeat business is to remove the friction of having to make a new decision every time. When a customer has to call, wait for a quote, compare options, and decide to book — every one of those steps is an opportunity for them to choose a competitor or simply not bother. When the arrangement is already in place, the decision has already been made.

For residential customers, tie packages to the calendar. A spring cleanout package and a fall cleanout package. A moving season package for homeowners who list their home in a predictably busy window. People already know these periods are stressful — arriving with a ready-made solution positions you as the company that anticipated their need rather than the one they had to think about hiring.

For property managers, a flat-fee annual plan that covers a set number of unit cleanouts removes the quote-every-time friction they find frustrating. They want one number, one call, and certainty that the job gets handled. If you can provide that, you've removed the main reason they'd ever shop around.

For contractors, volume arrangements work. They already build debris removal into their project bids — if you give them a reliable per-job rate and guaranteed availability, they have every reason to keep sending everything your way. A small discount on the tenth job in a month or guaranteed next-day dumpster delivery for regular clients costs you little and keeps them from testing a competitor.

The HVAC and plumbing industries figured this out years ago with membership plans — a small monthly fee buys priority scheduling, waived after-hours charges, and discounts on service calls. Customers love it because it removes uncertainty. You love it because it creates predictable revenue. The exact model doesn't translate directly to every service business, but the underlying principle does: customers will pay for certainty, priority, and convenience. If you can package those things, you've changed the relationship from transactional to structural.

Price packages to protect your margins, not just to win the sale. A recurring arrangement that erodes margin over time is worse than individual jobs at full rate. The goal is customers who commit, pay predictably, and cost less to serve per job because of the efficiency that comes from established relationships and known workflows.

The Personal Touches That Create Loyalty

Beyond formal packages and commercial relationships, a significant amount of repeat business is won through small gestures that cost almost nothing but create disproportionate goodwill.

The handwritten thank-you card is the most powerful of these and almost nobody does it. After every completed job, mail a card — a real physical card, not an email. Two or three sentences thanking them for the business and telling them you'd be glad to help again whenever they need it. Include a business card or a fridge magnet with your number. The card costs fifty cents. The magnet costs a dollar. Together they keep your name on their refrigerator until the next time the garage fills up or the basement gets out of control.

At Grizzly Junk Pros we built this into every job and the results were consistent — customers called back months or years later referencing the card, neighbors asked about the magnet, and the gesture alone generated a steady flow of repeat and referral business that we couldn't attribute to any other marketing activity. It worked because almost no competitor was doing it.

Seasonal check-ins by email or text keep you in front of past customers without being intrusive. Not promotional blasts — specific, useful messages timed to when the need is most likely to arise. A late-winter note reminding homeowners that spring is a natural time for a cleanout. A fall message about getting ahead of year-end projects. A post-holiday note about clearing out what didn't fit. These aren't aggressive sales emails — they're timely reminders from a company that's thinking about them, which is a different experience entirely.

Keep notes on your customers. In your CRM, tag every completed job with details that might matter later — the shed they mentioned wanting to clear next year, the contractor who typically calls monthly, the property manager who has twelve units and turns two or three every quarter. When you follow up six months later referencing something they told you in passing, it lands as genuine care rather than a sales call. Most customers are genuinely surprised when a service business remembers something specific about them because most service businesses don't bother.

Deliver an Experience Worth Coming Back To

Everything in this chapter — the packages, the commercial relationships, the personal touches, the follow-up sequences — only works if the underlying service is worth returning for. A customer who had a mediocre experience won't come back regardless of how many thank-you cards you send or how well-structured your follow-up sequence is.

The experience that creates repeat business is not complicated. Show up when you said you would. Do the job completely. Leave the space better than you found it. Be easy to communicate with before, during, and after. Follow through on whatever you committed to.

Research from Bain and Company found that 68% of customers who leave a service business do so because of perceived indifference — not because of price or a competitor, but because they felt the company didn't care about their business. That number is worth sitting with. The majority of the customers you're losing to attrition aren't being stolen by competitors. They're drifting away because nothing in the experience gave them a compelling reason to stay.

Every job is an opportunity to give them that reason. A crew that's professional and communicates clearly. A job site that's left clean. A follow-up text asking if everything looked good. These are small investments with large retention returns.

The customer who trusts you once, trusts you more the second time. By the third or fourth job, you've become their person — the default call when something needs to be dealt with. Getting to that status doesn't require anything extraordinary. It requires consistent execution of the basics, every time, at every job.

> *The first job wins a customer. Everything after that decides whether you keep them.*

Go Deeper

For more on this and everything else in the book, the Haulers' Edge Newsletter goes deeper every week. Scan the code or subscribe free at HaulingHubb.com.

PART 4: OPERATIONS & EFFICIENCY

CH. 19 RUN TIGHTER ROUTES, MAKE MORE MONEY

> *"By failing to prepare, you are preparing to fail." —Benjamin Franklin*

Every truck you send out costs money before it generates a single dollar. Wages start the moment the driver clocks in. Fuel burns from the first mile. Wear accumulates on the engine, the tires, the brakes — every mile, every day. By the time a truck pulls into the first job site, you've already spent real money just getting it there.

Most operators know this in theory. Few have actually calculated it. When I finally ran the numbers at Grizzly Junk Pros, it changed how I scheduled everything. A single trip — wages, fuel, maintenance factored in — was costing us around $50. Industry data puts the average operating cost of a service truck at $2.27 per mile *(American Transportation Research Institute, 2023)*. A 20-mile round trip is $45 in hard costs before the crew touches a single item.

Multiply that across every run, every day, every year — and then ask yourself how much you're leaving on the table by not treating routing like the profit lever it actually is.

Know Your True Cost Per Trip

Most operators don't know what a trip actually costs them. They know their fuel bill at the end of the month. They know their payroll. But they've never combined those numbers into a cost-per-trip figure that makes the daily routing decisions feel concrete.

Here's why that matters. When you know a trip costs $50,

shaving $10 off that trip through smarter routing doesn't sound like much. Multiply it by five trips a day, 250 working days a year, and that $10 improvement is worth $12,500 annually — from the same jobs, the same revenue, just smarter execution.

The two biggest components of trip cost are wages and fuel. Together they typically represent 60% or more of per-mile operating cost. Everything else — maintenance, depreciation, insurance — sits on top. The point isn't to obsess over the exact decimal. It's to internalize that every mile your trucks travel is money leaving your business, and not every one of those miles has to happen.

Small inefficiencies are surprisingly expensive when you run the math. Ten extra miles per day adds up to roughly $5,700 per year. Fifteen minutes of crew time wasted daily through poor scheduling or unnecessary waiting — across a year that's over 60 hours of lost productivity, which at a loaded labor rate of $50 per hour is $3,000 in recovered cost from a single scheduling habit. One unnecessary 30-mile round trip per week costs over $3,500 annually. None of these feel catastrophic in isolation. Together they slowly drain $10,000 or more from a small operation every year.

Route Smarter Before the Trucks Leave the Yard

Route sequencing is where the money is won or lost. The single highest-leverage change most service businesses can make is moving from "schedule jobs in the order they come in" to "schedule jobs by geography."

When you group jobs by location and sequence them in a logical loop, trucks spend more time doing paid work and less time driving between it. The driver who used to start downtown, drive to a suburb, come back to the city, and end the day in another suburb is now completing the same four jobs in a fraction of the drive time because they're sequenced geographically rather than chronologically.

In our dumpster rental operation the shift that made the biggest difference was moving from fixed appointment times to flexible windows. That gave our admin team room to build each day's schedule around geography — drop two dumpsters in the same neighborhood back to back, pick up a nearby rental in between, sequence dump runs to hit the transfer station at the right point in the route rather than as a separate trip. Drivers spend less time zigzagging empty across town. More jobs get done per day. Some days clever scheduling eliminates an entire trip back to the yard.

Transportation research consistently shows that poor routing adds 10% or more unnecessary miles to a typical service operation, while basic route optimization — whether through software or disciplined manual planning — reduces drive time and fuel consumption by 20–30% *(American Transportation Research Institute, 2023)*. That's not a marginal gain. On a fleet running five trucks, that's equivalent to eliminating one truck's fuel cost entirely.

You don't need expensive software to start. A free mapping tool with multiple stop optimization — Google Maps handles up to ten stops, purpose-built apps like OptimoRoute or Route4Me handle larger operations — is enough to see immediate improvement. The discipline of planning tomorrow's routes before the day ends is more important than the sophistication of the tool you use.

The UPS Right-Turn Lesson

UPS's routing research became a famous business case study for good reason. They discovered that eliminating left turns — which require waiting through traffic signals and create accident risk — from their delivery routes produced results that seem disproportionate to such a simple change: roughly 10 million gallons of fuel saved annually, 350,000 more packages delivered per year, and 28.5 million miles cut from their total route distance, enabling them to operate with over 1,000 fewer trucks *(UPS Pressroom)*. Safer, faster, and dramatically cheaper — from one routing principle applied consistently.

The principle scales directly to a small service operation. Not literally banning left turns, but developing a routing mindset that asks: where are we wasting motion? What backtracking are we accepting that we don't have to? What scheduling habits are costing us miles that better planning would eliminate?

At Grizzly Junk Pros we don't have a formal policy against left turns. But we plan routes to minimize bottlenecks, avoid school zones at dismissal time, and sequence stops so the truck is always moving forward rather than doubling back. The driver finishes the day with fewer miles on the truck, less fatigue, and more completed jobs. Those things compound.

A single truck idling wastes roughly 0.8 gallons of fuel per hour *(U.S. Department of Energy, 2023)*. It sounds minor until you track idle time across a fleet for a year. Cut idling through better scheduling — arriving at job sites when customers are ready, sequencing dump runs to minimize wait time at transfer stations — and the fuel savings are real and measurable.

Efficiency isn't one big breakthrough. It's a hundred small routing decisions made consistently, every day. UPS proved it at global scale. The same math works on a two-truck operation in your state.

Standardization: The Hidden Efficiency Multiplier

Route optimization and scheduling discipline produce significant savings. Standardization of equipment multiplies those savings by reducing the complexity that scheduling has to manage in the first place.

We covered standardization as an operational principle in Chapter 4. The logistics dimension is worth making explicit here because it's where the efficiency gains become most visible. When every dumpster in your fleet is the same size and every truck uses the same hitch system, dispatching becomes frictionless. Any container can go to any job. Any driver can operate any truck. There are no same-day calls to swap equipment because the customer's job required a different size than what's on the truck — because there is no different size.

We built the dumpster operation around 20-yard containers from day one — one size, one price structure, no exceptions. That decision wasn't the result of a painful lesson learned from running multiple sizes. It was an intentional choice made at the start, and watching other operators struggle with the complexity of managing multiple container sizes confirmed it was the right one. Competitors were repositioning equipment mid-day, sending second trucks because the first had the wrong size on it, and fielding customer calls about whether they'd ordered enough capacity. None of that existed in our operation because the decision was made before the first container ever hit a driveway. Starting with standardization is always easier than arriving at it after you've already built complexity into your system. If you're early enough in your growth to make this choice now, make it. If you're already running multiple sizes, the math on simplifying is worth running before you add another unit to the fleet.

Henry Ford's insight with the Model T wasn't stubbornness — it was efficiency. By eliminating variation in production he reduced cost and accelerated output at the same time. A service fleet operates on the same logic. Fewer equipment variations mean fewer scheduling errors, faster driver onboarding, and lower parts inventory for your mechanic. Each of those savings is modest on its own. Compounded across a full year of operations they show up meaningfully in your margin.

The customer experience benefits in ways most owners don't anticipate. We sell 10, 15, 20, 30, and 40 cubic yard options — customers still want to choose a size and a price point that fits their budget and their project. But every single one of those rentals delivers the same physical container: a 20-yard dumpster. We're transparent about this when customers ask — they're renting a price tier and a weight allowance, not a specific box size, and the physical container they get is a 20-yard. The customer who rents a 10-yard pays the 10-yard price and gets a 20-yard container sitting in their driveway. That extra capacity is the selling point. Most customers underestimate how much they're getting rid of — it happens on almost every job. With our model, when they inevitably fill past what they thought they'd need, the space is already there. They don't need a swap, they don't need a second delivery, they just keep loading. If they go over the weight allowance for their chosen size they pay for the additional weight. That's the only variable. Everything else is simple.

For customers who need what a 30 or 40-yard dumpster would typically provide, we offer two 20-yard containers. We can deliver both at the same time and place them at different areas of the job site, or deliver one and swap it out when it's full at no additional delivery charge. Two 20-yard containers take up less combined footprint than a single 40-yard dumpster and give the customer more flexibility in how they use the space. Our pricing for this option is elevated compared to what a competitor with an actual 30 or 40-yard container might charge — we're transparent about that — but we frame everything around what the customer gains rather than what it costs. More flexibility. Less footprint. No surprise fees. Faster swap when they need it.

We don't match competitor pricing on the larger size tiers. We don't need to. The model sells itself because it removes the friction that every other dumpster rental creates — the wrong size, the second delivery charge, the cramped driveway, the project that outgrew the container. We turned standardization into a customer benefit and priced it accordingly. Simplicity that serves your operations also serves your customer. That's the compounding effect of a good standardization decision — it pays you in efficiency and it pays you again in the customer experience.

Plan Tomorrow Before Today Ends

The operators who run the tightest routes share one habit: they plan tomorrow's schedule before the current day ends.

When tomorrow's jobs get sequenced the night before or at the end of the afternoon, the morning starts with a clear plan rather than a reactive scramble. Drivers know their first stop before they leave the yard. Admin knows the day's geographic groupings before the first customer calls with a scheduling question. There's no wasted time at 7am figuring out who goes where.

This planning time feels like overhead until you calculate what disorganized mornings actually cost. Fifteen minutes of delay for a two-person crew at a loaded cost of $50 per hour is $12.50. Multiply by 250 working days and that morning disorganization is costing $3,125 per year in labor alone, before you add the fuel and mileage cost of suboptimal routing that results from schedules assembled on the fly.

The afternoon planning session doesn't need to be long. Thirty minutes spent grouping tomorrow's jobs by geography, sequencing the route to minimize backtracking, identifying where dump runs fit into the day's flow, and flagging any jobs that need special scheduling attention pays for itself immediately. Build it into your operations as a non-negotiable part of closing the day.

If you have a dispatcher or admin handling scheduling, give them explicit authority to suggest schedule adjustments to customers when the adjustment improves routing efficiency. Most customers with flexible timing will accept a two-hour shift in their window if you're upfront about it — "I can get to you at 10 instead of 8 and you'd be helping us run more efficiently, which keeps our prices stable." That conversation almost always works and the routing benefit is immediate.

The Long Game on Efficiency

Every efficiency improvement in this chapter produces returns that compound over time, not overnight. The extra time spent planning routes, the investment in standardized equipment, the discipline of tracking cost per trip — all of these require upfront effort or capital that pays back across months and years.

This is where a lot of operators hesitate. The fuel savings from better routing don't show up until next month. The payback on standardized equipment doesn't materialize until next year. The temptation is to skip the planning, keep the mixed fleet, and get through today's jobs however they get done.

That short-term thinking is expensive. The operator who spends an extra thirty minutes planning routes today saves hours of wasted driving next week. The one who invests in a standardized fleet this year operates with meaningfully lower cost per job by next year. The compounding is real and it moves faster than most people expect.

Be patient with the big picture. Be relentless with today's work. Those two things aren't in conflict — they're the whole game. That framing applies directly here. The macro outcome is an operation that runs leaner and generates more profit from the same revenue. The micro execution is the daily discipline of planning routes, eliminating unnecessary trips, and making small efficiency decisions consistently.

Within a year of tightening our logistics at Grizzly Junk Pros, we were seeing the results in fuel costs, driver fatigue levels, and schedule consistency. By year two the operational savings were significant enough to affect our pricing flexibility — we could hold rates while competitors who hadn't tightened their operations were forced to raise theirs to cover increasing costs.

Efficiency isn't a one-time project. It's a mindset that becomes a competitive advantage over time as it compounds.

Two metrics tell you whether the work in this chapter is actually showing up in your numbers. Revenue per crew hour — total weekly revenue divided by total billable field hours — is the operational efficiency number that confirms your routing and scheduling improvements are producing real output. Set your baseline when the operation is running well and watch it weekly.

The second is asset utilization rate — the percentage of your trucks and equipment actively generating revenue versus sitting idle. For a dumpster operation that's dumpsters on job sites divided by total dumpsters in your fleet. For a truck-based operation it's billable truck days divided by available truck days. This number does more than measure efficiency — it tells you whether you're ready to grow. Sustained high utilization over multiple weeks signals genuine capacity pressure and justifies adding assets. A spike during peak season followed by a drop in slow months is normal and does not signal the need for more capital. Don't let one busy month drive a decision that commits you to costs you'll carry all year. Both metrics live in your weekly scorecard in Appendix A.

Every mile has a price tag. Every minute has a cost. Run tighter and watch the savings stack.

Go Deeper

For more on this and everything else in the book, the Haulers' Edge Newsletter goes deeper every week. Scan the code or subscribe free at HaulingHubb.com.

CH. 20 SOPS — GETTING OUT OF EVERY JOB

"Systems run the business and people run the systems."
— Michael Gerber

There's a version of your business that runs without you on every job. Where the crew knows exactly what to do when they arrive, how to handle the unexpected, what the job site should look like when they leave, and how to represent your company in a way that generates five-star reviews without you standing over them.

And there's the version most operators actually have — where everything depends on the owner being present, involved, and available. Where quality varies by who's on the crew that day. Where a new hire takes months to reach acceptable performance because the training is whatever you remember to tell them. Where the business is essentially a job you own rather than a company you've built.

The difference between those two versions is documentation. Specifically, Standard Operating Procedures — SOPs — the written processes that capture how your business does what it does, so anyone you hire can execute to the same standard you would execute to yourself.

This chapter is about building that documentation and what it actually buys you.

Why Most Service Businesses Never Build SOPs

The reason operators don't document their processes isn't laziness. It's a timing problem. When you're small and doing most of the work yourself, SOPs feel unnecessary — you know how everything works. When you're busy enough to need them, you're too busy to write them. So they never get written, and the business stays dependent on the owner indefinitely.

The cost of that dependency grows with every month you don't fix it. You can't take a day off without things going sideways. You can't hire confidently because training is inconsistent and results are unpredictable. You can't expand to a second truck or a second market without stretching yourself thin across everything. Every growth move you make adds complexity that lands on you personally because there's no documented system for anyone else to follow.

There's a well-known distinction between working in your business and working on it. Most spend their entire career in the first category and never touch the second. As Gerber argues in "The E-Myth", most small business owners are technicians who are good at the work itself, so they start a business doing that work. But a business requires systems, not just skilled execution. Without systems, you haven't built a business — you've bought yourself a job with extra responsibility.

SOPs are how you start working on the business instead of just in it. They're the foundation of everything else this book has covered — the consistent customer experience that drives reviews, the operational efficiency that drives margin, the team performance that eventually lets you step back.

What an SOP Actually Is

An SOP is a written description of how a specific task or process gets done in your business — detailed enough that someone who has never done it before could follow it and produce an acceptable result.

That's the test. Not "could I follow this" — you already know how. "Could someone I just hired follow this and do it right?"

Good SOPs are specific, sequential, and practical. They don't describe goals or values — they describe actions. Not "treat the customer with respect" — that's a value. "Introduce yourself by name when the customer answers the door, confirm the job scope before any items are moved, and ask if there's anything they want to double-check before you start" — that's an SOP.

They should be short enough to be usable. An SOP that takes forty-five minutes to read won't get read. A one-page checklist that takes five minutes to review before a job gets used every time.

And they should be living documents. When your process improves — when you figure out a better way to load a truck, handle an objection, or wrap up a job site — the SOP gets updated. The goal isn't to document how things were done. It's to document how things should be done right now, with the understanding that right now will evolve.

The SOPs Every Service Business Needs First

You don't need to document everything on day one. Build the SOPs that have the most leverage first — the ones that protect customer experience, protect your margin, and protect your reputation.

Job arrival and introduction. What happens in the first two minutes of every job. How the crew introduces themselves, how they confirm the scope of work with the customer before starting, how they protect the property — floor protection, door frame covers, pathway clearing. This is the impression that determines the review. Customers form their opinion of your company in the first five minutes. That impression shouldn't vary based on who showed up that day.

Job execution by service type. A separate SOP for each of your core services — residential junk removal, dumpster delivery and pickup, demolition work, estate cleanouts. What order do tasks happen in, how do items get sorted for disposal versus donation versus recycling, how is the truck loaded for maximum efficiency and minimum trips, what are the safety considerations specific to that job type. This is where your operational knowledge gets transferred from your head into a document anyone can follow.

Job completion and site walkthrough. What the job site should look like before the crew leaves. A specific checklist — sweep the area, check for any items the customer wanted to keep that might have been moved, do a walkthrough with the customer to confirm everything looks right, ask if there's anything else they need. This step is where five-star reviews are made or lost. A crew that completes a thorough walkthrough with the customer and asks "is there anything else we can do for you?" generates dramatically more positive reviews than one that finishes loading and drives away.

Customer communication during the job. How to handle a customer who wants to add scope mid-job, how to quote additional work on-site without undercutting your margin, how to handle a situation where something unexpected makes the job harder or more expensive than quoted. These conversations happen constantly. A crew that knows what to say handles them smoothly and professionally. One that's winging it creates awkward situations that cost you money or goodwill.

End-of-day truck and equipment checklist. What gets checked, cleaned, and logged before the truck comes back to the yard. Fuel level, equipment condition, any damage noted, supplies restocked. A crew that follows this consistently means you start every morning with trucks that are ready rather than discovering a problem at 7am.

Phone and booking process. We covered the call script in

Chapter 16. That script is an SOP. Write it down, train everyone who answers phones on it, and post it near every phone. A consistent booking process produces consistent conversion rates. Without it, results depend entirely on whoever happened to answer that day.

Review request process. How and when the review request goes out, what the message says, what platform to direct customers to, how to handle a customer who seems hesitant. This is one of the highest-leverage SOPs in the business — a systematic review request after every job compounds into a review profile that drives significant inbound leads over time.

How to Write an SOP Without It Taking Forever

The fastest way to document your processes is to record yourself or your best crew member doing the work, then transcribe and edit it into a written procedure. Talk through what you're doing while you do it. Narrate the decisions you're making, the things you're checking, the sequence you follow. A fifteen-minute video walkthrough of a job arrival and completion process becomes a one-page SOP in thirty minutes of editing.

Alternatively, walk through the process from memory in a voice memo during your commute. What happens first, what happens next, what are the things that trip people up, what does good look like versus acceptable. Transcribe it, organize it chronologically, cut anything that's obvious, and you have a working first draft.

Don't aim for perfect. Aim for usable. A rough SOP that gets used is infinitely more valuable than a polished one that never gets finished. The first version will have gaps — you'll discover them when someone follows it and encounters something you didn't document. That feedback makes the next version better. This is the iteration process that eventually produces a training system your business can rely on.

Format matters for usability. A numbered checklist works better than paragraphs for process documentation. Short sentences. Specific actions. Where there's a decision point — "if the customer wants to add items, do X; if the items are a different type than quoted, do Y" — make it explicit rather than leaving it to judgment. The whole point is to reduce the variability that comes from individual judgment on tasks that should be standardized.

SOPs and Hiring: The Connection Most Operators Miss

Here's the thing most operators don't connect: your ability to hire well is directly proportional to the quality of your SOPs. The reason hiring feels risky and training feels exhausting is that without documented processes, every new hire is learning a different version of how things work depending on who trains them and what that person remembers to mention.

With strong SOPs, hiring changes. The onboarding process becomes consistent — every new hire goes through the same documentation, the same demonstrations, the same checkpoints. Performance expectations are explicit from day one rather than assumed. Training time drops because the knowledge doesn't have to transfer through conversation — it lives in documents the new hire can reference.

It also changes accountability. When the SOP says the site walkthrough happens before the crew leaves and a customer reports it didn't happen, the conversation is specific and concrete. Not "you need to do better" but "this step was skipped — here's why it matters and here's what I need to see going forward." That specificity makes managing people significantly more effective.

Chapter 22 covers hiring in detail. When you get there, having your SOPs already built makes the hiring process dramatically cleaner. You can hand a candidate your job arrival and execution SOP during the interview and watch how they engage with it. Someone who reads it carefully and asks smart questions is a different hire than someone who skims it and says it looks fine.

SOPs and Your Exit — The Bigger Picture

Every SOP you write increases the value of your business. This connects directly to Chapter 29: Build It Like You're Going to Sell It.

A buyer evaluating your business is essentially buying your operation's ability to generate revenue without you personally running every job. Documented processes are proof that the operation can continue under new ownership. A business with strong SOPs, trained staff, and systematic processes commands a significantly higher valuation than an identical business where everything lives in the owner's head.

Even if you never plan to sell, the exercise of documenting your processes forces a useful discipline — you have to think clearly about how your business actually operates, which surfaces inefficiencies you'd otherwise never notice. The SOP writing process is itself an operational audit.

The business that runs without you isn't built through hiring alone or technology alone. It's built through the combination of documented processes, tools that execute those processes consistently, and people trained to follow both. SOPs are the foundation. Everything else builds on them.

Why SOPs Fail — And How to Make Them Stick

Most operators who've tried to document their processes have watched their crew ignore the documents. That's not a crew problem. It's a systems problem — and it has a specific solution.

SOPs fail for three reasons. First, the crew had no input in writing them, so they feel like rules handed down rather than agreements they helped shape. The fix is simple: record yourself or your best crew member doing the task, transcribe it, then review it with the people who will follow it. Their corrections make the SOP more accurate and their involvement makes them more likely to follow it. Second, there's no accountability mechanism — the SOP exists but nobody checks whether it's being followed. Spot checks during the first month a new SOP is live, brief debriefs after jobs where the process matters, and specific feedback when steps get skipped all close this gap. Third, SOPs go stale. A process written six months ago may not reflect how the job actually runs today. Build a quarterly review into your calendar where crew members flag what's outdated. The SOP that gets updated based on field feedback is the one the crew trusts.

The incentive structure we'll cover in Chapter 23 connects here directly. When crew performance reviews reference SOP compliance — and when the incentive program rewards consistent execution rather than just output — following documented processes becomes part of how people earn more. That connection, made explicit, is what moves SOPs from documents people know exist to behaviors people actually exhibit.

Start This Week, Not Someday

The reason "someday I'll document everything" never happens is that it doesn't have a deadline and it isn't urgent in the way that today's jobs are urgent. The fix is to make it small and immediate.

This week, write one SOP. Pick the process that, if it went wrong, would hurt your reputation or your margin most — probably the job completion walkthrough or the phone booking script. Write it in whatever rough format captures the steps. Share it with whoever is doing that task. Ask for feedback. Update it.

Next week, write another one. In two months you'll have eight to ten SOPs covering the core of your operation. In six months you'll have a training system. In a year you'll have documentation thorough enough that you can hire a new crew member and trust that what gets delivered to the customer on their first day on the job reflects your standards, not a guess.

That's the business that gives you back your time. Not a different business — yours, just systematized.

> *You can't scale what only exists in your head. Write it down.*

Go Deeper

For more on this and everything else in the book, the Haulers' Edge Newsletter goes deeper every week. Scan the code or subscribe free at HaulingHubb.com.

CH. 21 THE TOOLS THAT BUY BACK YOUR TIME

> *"The secret of getting ahead is getting started."* — *Mark Twain*

Here's what it looks like when it's running. Tuesday morning: you open your laptop before the crew leaves the yard. Seven leads came in overnight — five captured by the voice agent, two from the website form. Three follow-up sequences closed while you were asleep. The route for the day is already optimized. Two review requests went out from yesterday's jobs. You didn't touch any of it. That's the business you're building one tool at a time.

This chapter is about building that system — one tool at a time, starting with your biggest problem.

Every hour you spend on administrative work is an hour you're not spending on the work that actually grows your business. Scheduling jobs manually, chasing payments, following up on quotes from memory, tracking expenses in a shoebox, trying to remember who needs a callback — all of it is real work that consumes real time. And unlike the work that generates revenue, none of it gets better as your business grows. It gets worse.

Technology doesn't replace good operations. But the right tools eliminate the low-value repetitive work that currently fills hours of your week, reduce the human error that costs you money and customers, and create systems that keep running whether you're on a job, in a meeting, or on the beach in Florida.

Start With One Problem,

Not Five Platforms

The most common technology mistake service business owners make is trying to implement everything at once. They sign up for three platforms, spend a week watching tutorials, get overwhelmed, and end up using none of them effectively while paying subscriptions on all of them.

Start with your biggest pain point. What is the single thing in your operation right now that costs you the most time, creates the most errors, or loses you the most money? If you're already on a CRM platform, this chapter still applies. Every capability covered here exists in some form across the major platforms. Where I reference Service Hubb AI specifically, it's because that's what we built and run at Grizzly Junk Pros. If you're on a different platform, the principle is identical — the question is whether your current tool has the feature enabled and configured, not whether to switch. For most service businesses it's one of three things — leads falling through the cracks, scheduling chaos, or payments that are slow and inconsistent. Identify yours. Solve that first. When it runs without your daily attention, move to the next problem.

The adoption sequence that works for most service businesses: CRM and pipeline management first, then payment processing, then scheduling and dispatch, then route optimization, then expense tracking and accounting, then automation layered on top of all of it. Each tool builds on the foundation the previous one created.

What to Look for in a CRM

Every capability covered in this book — the follow-up sequences from Chapter 17, the AI voice and text agents from Chapter 16, the customer segmentation from Chapter 8, the retention automation from Chapters 11 and 18, the lead capture from Chapter 13, and the pipeline tracking from Chapter

14 — lives in your CRM. The platform you choose is where all of it runs. Here's what the right one needs to do.

Lead capture. Every inbound inquiry — web form, phone call, text message, AI voice agent interaction — should create a lead record automatically. No manual data entry, no sticky notes, nothing slips through. The lead is in the system the moment it arrives with the customer's name, contact information, and whatever details came in with the inquiry. A CRM that requires your team to manually log leads is one that will have gaps.

Pipeline stages. The system should let you build a pipeline that matches your actual sales process. For a service business that typically looks like: New Lead → Estimate Sent → Follow-Up → Booked → Job Complete → Review Requested. At each stage transition, an automation should fire — the estimate confirmation goes out, the follow-up sequence starts, the review request sends after the job closes. None of this should require manual action once it's configured.

Follow-up sequences. As covered in Chapter 17, the three-touch follow-up sequence — day one or two confirmation, day five to seven value touch, day thirty door-open message — should start automatically when an estimate is sent and pause automatically when the customer responds. The platform does the tracking. You do the closing.

AI voice and text agents. The better CRM platforms now include AI voice and text agents that answer overflow and after-hours calls, capture lead information, and route it directly into your pipeline. Look for a platform where the AI agent integration is native, not bolted on — and where the transparency principle covered in Chapter 16 is built in (the agent identifies itself as AI, not a human).

Customer history and segmentation. Every job, communication, and note should live on the customer record and be searchable. Tag customers by type — residential, contractor, property manager — and filter by tag when you run campaigns. When a customer calls back a year later, you should be able to pull their history in thirty seconds.

Retention automation. Seasonal campaigns for each customer segment should be buildable once and run indefinitely. The late-winter residential reminder, the quarterly contractor check-in, the post-holiday message — set and forget.

Reporting. The platform should tell you which services are converting best, which customer types are driving the most revenue, and which marketing channels are producing the most qualified leads — without you assembling it from memory and spreadsheets.

At Grizzly Junk Pros we run Service Hubb AI, which we built specifically for service businesses and handles all of the above in one integrated system. Jobber, Housecall Pro, and GoHighLevel are the other platforms most commonly used in the service trades — each covers the core functions, with varying depth on AI agent integration and automation flexibility. The right choice depends on your volume, budget, and how much of the automation you actually plan to build out. Any of them is better than no CRM. The worst system is the one you have in your head.

Payment Processing: Get Paid the Day You Do the Work

The standard service business payment cycle — complete the job, send an invoice, wait for a check, deposit it, wait for it to clear — is a cash flow problem masquerading as a business process. As we covered in Chapter 2, profit on paper means nothing if the cash isn't in your account to pay your bills. Every day between completing a job and receiving payment is a day your money is working for someone else.

Mobile card readers and payment links eliminate that cycle. The crew closes the job, the customer pays on-site or through a link on their phone, and the money is in your account within one to two business days. No invoice lag, no check in the mail, no collections calls.

At Grizzly Junk Pros we stopped accepting cash or checks entirely. Cards only. It keeps the books clean, eliminates the handling risk of cash, and means payment is never a conversation we're having two weeks after the job. Card processing fees — typically 2.7–3.5% — are a cost of doing business. Build them into your pricing and stop thinking about them as an expense. The trade-off is immediate payment, automatic record-keeping, and zero collections friction. It's worth it.

Integrate your payment processing with your accounting software — QuickBooks or Wave for most small service businesses. When payments connect to your books automatically, your financial picture stays current in real time rather than being something you reconstruct at the end of the month. The P&L and cash flow statements we covered in Chapters 1 and 2 are only useful if the data feeding them is current and accurate. Payment integration makes that happen without manual bookkeeping.

Square and Stripe are the most widely used payment platforms for small service businesses. Both offer mobile card readers, payment links, invoicing, and accounting integrations. Either works. The specific platform matters less than having one and using it consistently on every job.

Scheduling and Dispatch: The Heartbeat of Your Operation

Scheduling software confirms appointments automatically, sends customer reminders before the job, lets customers reschedule without a phone call, and keeps the day's jobs organized in a view your team can see and update in real time.

For a solo operator or small team, a shared Google Calendar with job details noted in each event is a functional starting point. It costs nothing and keeps everyone looking at the same schedule.

As volume grows, purpose-built field service management software handles more complexity — crew assignment, job notes, customer communication, GPS tracking, and invoice generation all from one platform. Jobber, Housecall Pro, and Service Fusion are the most commonly used platforms in the service trades. Service Hubb AI also handles scheduling and dispatch as part of its integrated system, which means jobs flow from the CRM pipeline directly into the dispatch calendar without re-entering information.

The specific scheduling feature that pays for itself fastest is automated customer reminders. A text the day before and the morning of a job reduces no-shows and last-minute cancellations significantly — research puts the reduction at 30–40% for service businesses with automated reminder systems *(Software Advice, 2023)*. A no-show costs you the full trip cost covered in Chapter 19. A reminder that prevents one no-show per week is worth hundreds of dollars a month in recovered operational cost.

Route Optimization: Technology for the Road

We covered routing strategy in Chapter 19. The technology that executes that strategy is worth naming specifically.

For a one or two truck operation, Google Maps handles multi-stop route optimization adequately. Input the day's jobs in the morning, hit optimize, and the sequence is done. Free and fast.

For larger operations or anyone serious about maximizing daily job capacity, dedicated routing software produces measurably better results. OptimoRoute and Route4Me both integrate your job list, service windows, truck capacity, and dump run requirements into optimized daily routes that reduce total drive time and mileage. The time savings at scale — across a fleet running fifty or more jobs per day — is significant enough to justify the monthly cost many times over.

Fleet tracking and telematics — GPS devices on every truck — give you real-time visibility into where your vehicles are, how long they've been at each stop, and whether drivers are following optimized routes. For multi-truck operations this isn't a surveillance tool, it's a dispatch tool. When a job runs long, you see it in real time and can adjust the afternoon's schedule before it creates a cascade of delays. Samsara and Verizon Connect are the most widely used fleet tracking platforms in small to medium service businesses.

Expense Tracking and Financial Visibility

Every dollar your business spends that doesn't get tracked is a dollar that doesn't show up in your P&L, which means your financial picture is inaccurate, your tax deductions are incomplete, and you can't see the margin leaks that are quietly eroding your profitability.

Expense tracking apps — QuickBooks, Wave, or Expensify for expense-specific tracking — connect to your business bank account and credit card and categorize transactions automatically. Fuel, dump fees, supplies, equipment repairs — everything gets logged without manual entry. Your P&L updates in real time. When you do your quarterly 80/20 audit or your monthly margin review, you're working from current data rather than reconstructing the past.

The business credit card is the simplest expense tracking tool available. Every business expense goes on the card, the statement provides a complete monthly record, and the integration with accounting software means nothing requires manual input. If you're still running business expenses through a personal account or paying cash for anything business-related, that's the first thing to fix. The separation isn't just about tracking — it's about the legal and tax protection of keeping business and personal finances completely distinct, as covered in Chapter 6.

Mileage tracking is worth automating separately. Apps like MileIQ or Everlance run in the background on your phone and automatically log business miles. At the current IRS standard mileage rate (check IRS.gov for the current year's figure), a service business driving 20,000 business miles annually has a $13,400 deduction sitting in a mileage log that goes uncaptured in most service businesses.

Communication Tools: Keep the Team Aligned

Miscommunication between team members is one of the most consistent sources of operational errors in growing service businesses — double-booked jobs, wrong addresses, missing job notes, crew showing up unprepared for what a job requires.

Internal communication tools keep everyone working from the same information. For small teams, a dedicated group text thread or WhatsApp group for daily operational coordination works fine. For larger teams or anyone who needs more structure, Slack provides channels by topic — one for dispatch updates, one for job issues, one for general team communication — with searchable history that prevents "I didn't see the message" problems.

The job notes feature in your scheduling or CRM platform matters as much as any standalone communication tool. When a customer mentions a gate code, a fragile item, a neighbor's parking concerns, or an access limitation during booking, that detail needs to travel from the booking conversation to the crew before they arrive. A crew that shows up with the right information handles the job better and creates a better customer experience than one that's discovering surprises on-site.

Customer-facing communication — automated confirmations, day-before reminders, on-my-way texts — builds the professional impression that generates reviews and repeat business. Most of this runs automatically through your CRM once configured. The crew sending a "heading your way, about 15 minutes out" text before every arrival isn't a big gesture. But it consistently shows up in five-star reviews because so few competitors do it.

The Adoption Sequence in Practice

Everything in this chapter connects and compounds. Service Hubb AI captures the lead and runs the follow-up. Payment processing gets you paid immediately. Scheduling keeps the day organized and reduces no-shows. Route optimization cuts trip costs. Expense tracking keeps the books accurate. Communication tools keep the team aligned. Together they create an operation that runs more reliably, more profitably, and more independently than one held together by your personal attention to every moving part.

But you don't build all of this in a week. The sequence matters.

Start with a CRM and payment processing — these two together capture every lead, follow up on every estimate, and get you paid on every job. That's the revenue foundation. Once those run cleanly, add scheduling with automated reminders.

Then route optimization. Then expense integration. Each addition builds on a stable foundation rather than adding complexity to a chaotic base.

The operators who adopt technology gradually and master each tool before adding the next beat the ones who try to do everything at once. Be patient with the learning curve. Most of these platforms have tutorials built in and support teams that exist specifically to help service business owners get set up. The first two weeks with any new tool are the hardest. By week four it's part of how your business runs.

The endgame is a business where the tools handle the repetitive work, the SOPs from Chapter 20 handle the execution standards, and your time is spent on the decisions and relationships that actually require you.

> *Every hour a tool saves you is an hour you earned back.*
> *Build the system and get your time back.*

Go Deeper

For more on this and everything else in the book, the Haulers' Edge Newsletter goes deeper every week. Scan the code or subscribe free at HaulingHubb.com.

CH. 22 YOUR FIRST HIRE WILL MAKE OR BREAK YOU

"Hire character. Train skill." — Peter Schutz

The moment you bring someone else into your business, everything changes.

Before the first hire, every problem is yours to solve and every result is yours to own. You control the quality, the customer interaction, the pace, the standard. It's exhausting, but it's controllable. The moment someone else is representing your business on a job site, you've introduced a variable you can't fully control — and that variable can build your reputation or destroy it depending on who you hired and how well you set them up.

Most service business owners make their first hire reactively. They're so overwhelmed with work that they hire the first person who seems capable and available. They put them in a truck, show them a few things, and send them out. Sometimes it works. Often it doesn't. And the cost of a bad first hire isn't just the time it takes to let them go and start over — it's the customers who had a bad experience with someone who shouldn't have been representing your company in the first place.

Your first hire is the most important business decision you'll make outside of your pricing and your service focus. Get it right and you've multiplied your capacity without compromising your standards. Get it wrong and you've added cost, liability, and headaches while creating a customer experience problem that shows up in your reviews before you realize what happened.

Before You Hire: Get the Foundation Ready

Hiring before you're ready to onboard someone properly is one of the most common and expensive mistakes in a growing service business. Ready doesn't mean perfect — it means you have the minimum infrastructure in place so the person you hire can actually succeed.

That minimum infrastructure is three things.

SOPs for the core job functions. Chapter 20 covered this in full. Before you hire, you need written documentation for how a job gets done — arrival, execution, completion walkthrough, customer interaction, truck end-of-day checklist. If you don't have these documented, you're hiring someone to improvise your standards, which means your standards become whatever that person decides they are on any given day.

A way to pay them correctly. This sounds obvious but is consistently underprepared for. You need a payroll system set up before the first paycheck is due — not the week before. Gusto, QuickBooks Payroll, and Heartland handle small business payroll at reasonable cost and ensure your payroll taxes, withholding, and filings are handled correctly. Mishandled payroll is both a legal and an operational problem. Fix it before it becomes one.

Workers' compensation insurance. In most states, workers' compensation is legally required the moment you have your first employee. In a physical service business — lifting, loading, operating equipment — the exposure is real. An employee injured on the job without workers' comp coverage creates a personal liability that can end your business. Get the coverage before the first day, not after.

If you're not ready on these three fronts, spend one more week getting them in place before you start the hiring process. The delay is worth it.

Employee or Subcontractor — Know the Difference

This is the question most first-time hirers get wrong, and the IRS has specific definitions that determine which one you're actually dealing with — regardless of what your paperwork says.

A subcontractor controls how they do their work, sets their own hours, provides their own equipment, and typically works for multiple clients. You pay them as a 1099 contractor, they handle their own taxes, and you have no obligation for benefits or workers' comp under their arrangement.

An employee works on your schedule, uses your equipment, follows your processes, and works primarily for you. If that describes the person you're bringing on — and it usually does for a service business crew member — they're legally an employee regardless of what you call them or what agreement you have them sign.

Misclassifying an employee as a contractor is one of the most common and costly legal mistakes in service businesses. The IRS can reclassify the relationship retroactively, assess back payroll taxes, interest, and penalties across every year of the misclassification. Several states have their own classification tests that are stricter than the federal standard. The risk is real and the consequences are severe.

If the person is doing your work, on your schedule, using your truck, following your SOPs — they're an employee. Set them up correctly from the start.

What to Look For Before You

Look at Experience

For a first hire in a service business, character matters more than experience. Experience can be taught. Work ethic, reliability, and how someone treats other people cannot be installed through training.

The traits that predict success in a service business employee are consistent across trades and markets. Reliability — they show up when they say they will, they communicate when something changes, they don't disappear. Physical work ethic — they move with urgency, they don't wait to be told what to do next, they take pride in finishing a job completely rather than stopping at good enough. Customer instinct — they're naturally decent with people, they read situations correctly, they don't create friction where there doesn't need to be any. And coachability — they take feedback without defensiveness, they want to improve, they ask questions rather than guessing.

Experience operating a truck, handling heavy items, or working in a service trade is useful but secondary. A person with strong character and no experience will outperform a person with weak character and years of experience within six months — because you can teach skills but you can't teach someone to care.

Where do you find people like this? Your best hires often come from people you've already observed. Former employees from other contexts — a kid who worked hard at a landscaping company you know, someone who impressed you as a server or retail worker, a referral from a current employee or contractor whose judgment you trust. Trade school programs and workforce development organizations are underutilized sources for motivated young workers who want to learn a trade. Referrals from current employees are consistently the highest-quality source across service industries — someone your best worker is willing to vouch for comes pre-screened for culture fit.

Job boards work but require more filtering. Be specific in the posting about what the job actually involves — physical demands, schedule, pay structure. Vague postings attract vague candidates. A posting that says "physically demanding outdoor work, lifting up to 75 pounds regularly, early starts, customer-facing" filters out anyone who isn't prepared for what the job actually is.

The Interview Isn't Just What They Say

The interview for a service business position is as much about observation as questions. How did they communicate about the interview itself — were they on time, did they confirm, did they follow up? How do they present themselves — not necessarily formally, but with basic cleanliness and respect for the interaction? How do they talk about previous employers — with maturity and accountability, or with excuses and blame?

Ask questions that reveal character rather than just competence. "Tell me about a time a job went sideways and how you handled it" shows you how they manage unexpected problems. "What's something you got better at in your last job" reveals self-awareness and growth orientation. "What would your last boss say your biggest weakness was" tests honesty — anyone who claims they have no weaknesses is either lying or lacks self-awareness.

Present your job arrival SOP from Chapter 20 during the interview. Ask them to walk you through how they'd execute it. Someone who reads it carefully, asks a clarifying question or two, and demonstrates they understand why each step matters is a better candidate than someone who skims it and says it looks simple. The SOP interaction tells you more about their professionalism than most interview questions will.

Do a working interview before making a final decision. Have

them shadow you or your best crew member on an actual job for half a day. Observe how they interact with the customer, whether they're reading the job environment and adapting, whether they're asking good questions or waiting to be told everything. A person who looks great in a conversation and falls apart in the field is a much better discovery before you hire them than after.

Background and reference checks are non-negotiable for anyone driving your vehicles or entering customers' homes. Most customers are trusting you with access to their property. Verifying a candidate's background and speaking to a previous employer is basic due diligence that protects your customers and your business. It also signals to the candidate that you run a serious operation with standards — which sets the right expectation before their first day.

One hiring practice that sounds counterintuitive until you've done it: bring two candidates in at the same time and train them together.

The instinct is to hire one person, see how they work out, and then hire the next one if the first doesn't stick. The problem with that approach is the math. If your first hire doesn't work out after 30 or 60 days, you're back at zero — reposting the job, screening again, scheduling interviews again, starting the training process from scratch. Do that two or three times and you've spent four to six months trying to fill one position while your operation is either understaffed or dependent entirely on you.

Training two people simultaneously compresses that timeline dramatically. Most service trucks already have three seats — a driver and two passenger seats. You're not creating a logistical problem. You're filling capacity that already exists. The additional cost is one extra daily rate during the training period. That cost is real but it's a fraction of what repeated failed hires cost you in time, operational disruption, and your own energy.

The other benefit is comparison. When you're training two people at the same time on the same jobs with the same customers, you see things you'd never catch evaluating them weeks apart. How they each handle a difficult customer interaction. How they respond when a job is harder than expected. Who asks better questions. Who shows initiative without being told. Who the customer gravitates toward. That parallel observation gives you information that sequential hiring simply can't produce.

Inevitably one of the two candidates doesn't work out — that's the assumption you're training toward, not hoping against. When it happens you already have someone else trained and ready. The position is filled. The operation keeps moving. You didn't lose a month finding out which hire was the right one.

The slightly elevated cost of training two people at once is the best insurance policy you can buy against the timeline cost of starting over.

The First 90 Days: Where Hires Succeed or Fail

The most expensive hiring mistake isn't choosing the wrong person — it's choosing the right person and failing to onboard them properly. A new hire who doesn't get clear direction, adequate training, and honest feedback in their first ninety days often fails not because they weren't capable but because they were set up to figure things out on their own.

Week one should be structured and supervised. They're working alongside you or your best crew member on every job. They're reading and reviewing every SOP. They're not doing independent work — they're learning the standard. This week is an investment that reduces errors and rework for the following years of the employment relationship.

Week two through four they start taking more responsibility with you or a senior crew member present. They're executing job arrival and completion processes independently while you observe and give real-time feedback. The feedback during this period needs to be specific and immediate — not a vague "good job" or "you need to do better" but exactly what was done right, exactly what needs to change, and exactly what you'll be watching for next time.

Days thirty through ninety is where you're evaluating whether this person meets your standard and whether they're growing. Are they improving? Are they showing initiative? Are they handling customer interactions in a way that generates the experience you want? Are there patterns in their performance that concern you?

The 90-day mark is a natural review point. Have an explicit conversation — here's what you've done well, here's what needs to continue improving, here's what I need to see to know this is working long-term. That conversation should not be a surprise to either party. If you've been giving consistent feedback, they already know where they stand. The formal review is confirmation, not revelation.

If someone isn't meeting the standard at 90 days despite clear feedback and genuine effort on both sides, that's important information. Some people aren't a fit regardless of character and intention. Acting on that information early is better for both parties than prolonging a relationship that isn't working.

Compensation That Attracts the Right People

Paying below-market rates to save money on labor is one of the most expensive decisions in a service business. Undercompensated employees perform at the level they're paid, leave when anything better comes along, and tell their network that your company isn't worth working for.

Know what your market pays for the role you're filling. Check current wage data for your market on the BLS website or Indeed — rates shift with inflation and local labor conditions. At the time of writing, most US markets paid service business crew members between $18 and $28 per hour depending on location, experience, and physical demands. Your local range will vary and trends toward higher over time *(Bureau of Labor Statistics)*. Physical labor trades have seen meaningful wage increases over the past several years as labor demand has outpaced supply.

Pay the market rate or slightly above it. The additional cost compared to underpaying is typically $5,000–$10,000 per year. The cost of high turnover — recruiting, background checks, onboarding time, the learning curve tax on productivity, and the customer experience degradation during that learning curve — is often 50–200% of their annual salary *(SHRM)*.

Beyond base pay, think about what else makes the total package competitive. Consistent scheduling — people value knowing their schedule. Paid time off for full-time employees, even if it starts modestly. Clear performance-based raises so the person can see a path to earning more as they develop. A workplace where they're treated with respect and their work is acknowledged.

The service business labor market is competitive. The operators who build reputations as good places to work — who treat employees fairly, pay on time, give clear feedback, and invest in their development — have dramatically lower turnover than those who don't. That reputation spreads through word of mouth in local labor markets the same way customer reviews spread online. Your next best hire might come from your current employee's recommendation because they're glad they took the job.

Managing Performance After the Hire

Hiring well is the start. Managing performance consistently is what determines whether a good hire stays good or drifts into mediocrity.

The most important management principle for a small service business is that expectations need to be explicit and feedback needs to be immediate. Vague expectations produce unpredictable performance. Delayed feedback produces confusion about what the standard actually is.

When something is done right — specifically right, not generically good — say so. “The way you handled that customer when she asked about the additional item was exactly right — you got the quote before agreeing to it instead of absorbing the extra work. That's what I want to see.” That specific acknowledgment reinforces the behavior you want repeated.

When something is done wrong, address it the same day. Not in front of the customer, not in front of the crew — privately, specifically, and constructively. “When you left without doing the walkthrough on the second job today, that's a step in our SOP for a reason. It's where we catch issues before the customer calls us about them. I need that to happen on every job.” Specific, tied to the SOP, explains the why, states the expectation clearly.

Document performance conversations. Not to build a legal case — to create a record that protects both you and the employee. If someone is let go and claims it was unfair, documented performance conversations demonstrate that feedback was given and standards were communicated. More often, documentation helps you track whether someone is improving over time — which is hard to see clearly without a record.

When It's Time to Let Someone Go

The most expensive employee in your business is the one you know isn't working but haven't let go yet. Every day they continue costs you in productivity, in quality delivered to customers, in the effect their performance has on your other employees, and in the mental energy you spend managing around the problem.

The threshold for letting someone go is a combination of performance and effort. Someone who isn't meeting standards but is genuinely trying and showing improvement deserves more time and support. Someone who isn't meeting standards and isn't putting in the effort to improve — who is coasting, creating problems, or undermining team culture — should be let go cleanly and without delay.

When you make that decision, be direct and respectful. The conversation should be short, clear, and private. State what you're doing and briefly why. Don't make it a negotiation or an extended performance review. Give them what they're owed — final paycheck, any accrued benefits per your agreement — and handle the separation professionally.

Most first-time managers delay this conversation far too long because it's uncomfortable. The discomfort of the conversation lasts an hour. The cost of delaying it lasts months.

One Good Hire Changes Everything

Done right, your first hire isn't just additional capacity. It's proof of concept that your business can run with someone other than you delivering it. That proof of concept is the foundation of every subsequent hire, every expansion, every move toward the business running without you needing to be on every job.

The operator who hires well, onboards thoroughly, manages clearly, and pays fairly will look back on their first hire as the moment the business started becoming something real — not just a job they owned, but a company they built.

> *Hire slowly. Onboard completely. Manage directly. That's how one good person becomes the first of many.*

Go Deeper

For more on this and everything else in the book, the Haulers' Edge Newsletter goes deeper every week. Scan the code or subscribe free at HaulingHubb.com.

CH. 23 BUILD A TEAM THAT DOESN'T NEED BABYSITTING

> *"Train people well enough so they can leave. Treat them well enough so they don't want to." —Richard Branson*

The goal isn't a team you manage. It's a team that manages itself.

That distinction sounds small but it changes everything about how you run your business. A team you manage requires your constant presence — you're the quality control, the motivation, the problem-solver, the backstop for every decision. A team that manages itself executes to your standard because that standard has been communicated clearly, reinforced consistently, and backed by a culture where people actually want to do good work.

Most service business owners never get there. They hire, they train loosely, they manage reactively, and they end up permanently in the weeds of their own operation — unable to step back because the moment they do, quality drops. If that sounds familiar, the problem usually isn't the people. It's the system around them.

Chapter 20 gave you the SOPs. Chapter 22 gave you the hiring foundation. This chapter is about what happens after the hire — how you develop people, how you motivate them, how you retain the good ones, and how you build a culture where your standard becomes their standard.

Train Beyond the Technical

Most service business training stops at the physical — how to lift safely, how to load the truck, how to operate the equipment. That's necessary but it's not sufficient. The crew member who knows how to do the job technically and doesn't know how to interact with a customer is a liability every time they're face-to-face with someone whose house they're working in.

Soft skills training is as important as technical training for a service business and it gets skipped almost everywhere. Specifically: how to greet a customer when you arrive, how to confirm the scope of work before starting, how to handle a question you don't know the answer to, how to recognize when a customer seems unhappy and what to do about it, and how to close out a job with a walkthrough that generates a five-star review instead of a complaint.

These aren't natural behaviors for everyone. They can be taught, practiced, and reinforced until they become automatic. The crew member who finishes a cleanout and says "let me do a quick walkthrough with you to make sure everything looks right before we head out" is executing a trained behavior. So is the one who, when a customer asks a question beyond their knowledge, says "I'm not sure on that one — let me have my boss give you a call this afternoon." Both of those interactions create trust. Neither happens consistently without training.

Upselling is another soft skill worth developing deliberately. The crew member who finishes the scheduled work and asks "is there anything else you'd like us to grab while we're here?" is generating incremental revenue with zero additional acquisition cost. We covered in Chapter 11 that the probability of selling to an existing customer is 60–70% compared to 5–20% for a new prospect *(Marketing Metrics, Paul Farris et al.)*. Your crew is having that conversation at the highest-trust moment in the customer relationship — right after completing good work. A crew trained to ask that question consistently turns single jobs into bigger tickets regularly.

Train on these skills the same way you train on technical

skills — with demonstration, practice, and feedback. Role-play the customer walkthrough. Do a mock upsell conversation. Debrief after jobs where something went particularly well or particularly poorly. The investment is small and the return on customer experience and revenue is disproportionate.

Turn Routing Into a Team Discipline

We covered route optimization as an operational strategy in Chapter 19. The team dimension of it is worth addressing separately — because efficient routing isn't just a dispatch function. It's a crew culture function.

When your crew understands why routes are planned the way they are — why jobs are grouped geographically, why the dump run is scheduled at a specific point in the day, why the afternoon sequence flows the way it does — they become active participants in efficiency rather than passive recipients of a schedule. They catch when a job sequence doesn't make sense and flag it. They suggest route adjustments based on traffic patterns they've learned from experience. They take ownership of getting through the day efficiently rather than just working through whatever list they were given.

Balance workloads deliberately. A crew that's consistently overloaded while another runs light doesn't just produce inconsistent output — it produces resentment. The crew running hard every day while another crew has slack builds a sense of unfairness that erodes morale faster than almost anything else. Use your scheduling data to distribute workload equitably, especially on the physically demanding days.

Build in realistic breaks. A service business crew doing physical labor in varying weather conditions needs recovery time built into the schedule, not squeezed out of it. A crew that's running ragged by 2pm makes mistakes, creates customer experience problems, and generates the kind of fatigue that leads to injuries. More jobs completed with a tired crew is not better than fewer jobs completed with a crew operating at full capacity. The math on safety incidents, quality errors, and turnover almost always favors the schedule that respects human limits.

Build an Incentive Structure That Pulls the Right Behaviors

People do more of what gets recognized and rewarded. That's not manipulation — it's just how motivation works. If the behaviors you need most from your team are never acknowledged or rewarded, you're relying on intrinsic motivation alone, which varies significantly by person and declines over time without reinforcement.

Research consistently shows that performance incentive programs improve output by 25–44% when structured correctly *(Incentive Research Foundation)*. The key phrase is structured correctly. Poorly designed incentives produce perverse behavior — a crew rushing through jobs to hit a volume metric, cutting corners on quality because speed is what gets rewarded, or competing internally in ways that damage team cohesion.

Design incentives around outcomes that matter to your business and your customers. Customer satisfaction ratings — tracked through your review score and post-job feedback — reward the quality of customer interaction, not just the volume of work. Upsell performance rewards the revenue-generating behavior you want to see. Safety records reward the protective behavior that keeps your costs and liability down. Attendance and reliability reward the foundation everything else is built on.

The rewards don't need to be large to be effective. A monthly bonus for the crew with the best review scores. A $50 reward for hitting a weekly upsell target. A half-day off for meeting a productivity goal as a team. An Employee of the Month recognition with a public acknowledgment in front of the team. Public recognition from you personally — "I want to point out how Marcus handled that difficult job on Tuesday" — is more motivating for many people than a cash bonus because it's specific, visible, and personal.

Keep the incentive structure simple enough that every crew member can tell you exactly what they need to do to earn it. Complexity kills engagement. If someone has to do mental gymnastics to figure out whether their performance qualifies for a reward, they've already disengaged from the program.

Let the team have input on what they find motivating. What you think they want and what they actually want are not always the same thing. A quick conversation — "what would make you feel recognized for a great week?" — often surfaces ideas that cost you little and matter to them significantly.

Pay Structure Is a Management Tool

How you pay your team shapes how they behave before any incentive program enters the picture. An hourly-only structure pays for time. A structure that ties compensation to outcomes pays for results. The difference compounds over years into two completely different cultures. When we layered performance incentives on top of our base pay structure at Grizzly Junk Pros, the shift wasn't dramatic overnight — but it was real. What changed was employee buy-in. The crew stopped being people executing a schedule and started being people with a stake in how efficiently the day ran. That buy-in shows up in the small decisions — the ones you're not there to make for them.

Hourly pay has one fundamental flaw for a service business — it removes the connection between effort and outcome. A crew member who works efficiently and delivers a great customer experience earns the same as one who drags the day out and does the minimum. There's no upside for performing well and no downside for performing slowly. You've built a system that rewards showing up, not producing.

The fix isn't eliminating hourly pay — it's layering performance incentives on top of a fair base. A per-job completion bonus that rewards efficiency. A quality component tied to review scores. A revenue share on upsells identified and closed on-site. These structures cost you money only when your team is making you money — which is the exact alignment you want.

Keep the structure simple. Complex incentive programs create confusion and resentment. One or two clear performance metrics with transparent tracking and consistent payout beats an elaborate points system nobody fully understands. The team member who knows exactly what they need to do to earn more will work toward it. The one navigating a complicated formula stops trying.

Review your pay structure the same way you review your pricing. Ask honestly — does it reward what I actually want from my team? If a crew member can do the minimum and collect the same check as someone who goes above and beyond, you've answered your own question. That's a systems problem, not a people problem. And systems problems have systems solutions.

Create the Feedback Culture Most Teams Never Have

The front line of your business sees things you don't. Your crew is on job sites every day, talking to customers, navigating the operational realities of the work in ways that surface information you'd have to deliberately create systems to capture. The equipment quirk that's going to become an expensive failure. The customer pattern that keeps coming up. The scheduling bottleneck that's adding twenty minutes to every afternoon. The approach that's working better than what the SOP currently describes.

Most of that information never reaches the person who can act on it because no mechanism exists for it to travel. Crew members don't speak up because nobody asked, or because the last time someone made a suggestion nothing happened, or because the culture signals that criticism of how things are done is unwelcome.

Build the mechanism deliberately. A brief weekly huddle — ten to fifteen minutes at the start or end of the week — where the agenda explicitly includes "what's working, what's not, what should change" gives feedback a structured home. It doesn't have to be formal. It just has to happen consistently.

The more important part is what you do with what you hear. When a crew member raises an idea or a concern, respond to it explicitly — either by implementing it, explaining why you're not implementing it, or committing to look into it and following up. The response to feedback matters more than the feedback mechanism itself. Gallup research consistently shows that employees who feel their voice matters at work demonstrate measurably higher engagement — and engagement shows up directly in the quality and consistency of what your team delivers to customers *(Gallup Workplace Research)*. That engagement shows up in quality, in customer interactions, and in whether they stay or leave.

When a crew member's suggestion gets implemented, acknowledge it publicly. "Marcus pointed out last week that we were losing time on the dump run sequence — we adjusted it and saved almost an hour yesterday. That kind of input is exactly what makes us better." That acknowledgment does two things: it rewards the behavior you want to see more of, and it signals to everyone else that speaking up has real value.

Tell Your Team Where the Business Is Going

Most crew members have no idea whether the business had a good month or a difficult one. They show up, do the work, and go home. The only signal they get is whether there are jobs on the schedule.

That information gap is a leadership problem. A team that doesn't understand the business context for what they're doing can't make good decisions. They can't prioritize intelligently. They can't identify problems you haven't seen yet. They can't contribute to solutions. They're executing tasks in a vacuum and wondering why you seem stressed sometimes.

You don't need to share every financial detail with every employee. But sharing the basics — where the business is, what the goals are, and how their work connects to those goals — changes how people show up.

A simple monthly five-minute team update covers the territory. How last month went in plain terms — strong, slow, or right on target. What the focus is for the coming month. Any process changes or priorities worth knowing about. Recognition for anyone who did something worth acknowledging publicly.

That five minutes of transparency does something no performance review can replicate. It makes your team feel like participants in a business rather than parts in a machine. Research from Gallup consistently shows that employees who understand their organization's goals and feel their work connects to them demonstrate significantly higher engagement and retention — and engagement shows up directly in the customer experience your team delivers every day.

The team that knows what winning looks like will help you get there. The one that's never been told has no way to aim.

Retention Is a Strategy, Not an Outcome

High turnover is expensive in ways most operators underestimate. Direct costs — recruiting, background checks, onboarding time — are visible. The indirect costs are larger: the productivity gap during the learning curve, the customer experience degradation while a new person finds their footing, the institutional knowledge that walks out the door, and the effect on remaining employees who are absorbing extra work while a replacement gets up to speed.

As covered in the last chapter, replacing an employee typically costs 50–200% of their annual salary when all costs are accounted for. Investing a fraction of that in retention produces a dramatically better return than repeatedly replacing people.

Retention is built on a few foundations that most owners either take for granted or neglect entirely.

Fair and competitive pay. We covered the wage benchmarks in Chapter 22. Paying below market to save on labor is a false economy. The turnover cost exceeds the payroll savings consistently. Pay people fairly and adjust rates as their performance and tenure justify it.

Clear growth paths. People stay longer when they can see where they're going. A laborer who knows that six months of strong performance puts them in line for crew lead, and that crew lead comes with a pay increase and more responsibility, has a reason to invest in their performance beyond just today's job. Even in a small operation with limited formal advancement, you can create growth through skill development, expanded responsibility, and compensation growth tied to tenure and performance.

Treating people like people. This sounds basic because it is.

Acknowledge birthdays and work anniversaries. Say thank you specifically and regularly. Celebrate wins as a team. Sponsor a crew lunch after a great week. Give people branded gear they're proud to wear. None of these things are expensive. All of them communicate that the people doing the work are valued, not just the output they produce.

Consistency and fairness in management. Nothing drives good employees out faster than inconsistent standards — where some people are held accountable and others aren't, where favoritism shapes decisions, where the rules that apply to one crew member don't apply to another. Your best employees have options. If the environment they're in doesn't feel fair, they'll find one that does.

The team that stays together gets better together. Every month a good crew member stays, they're more efficient, more skilled at customer interaction, more capable of handling unexpected situations, and more valuable to your operation. Retention compounds in exactly the way turnover costs compound — just in the opposite direction.

The Standard Becomes the Culture

Everything in this chapter — the training, the incentives, the feedback mechanisms, the retention investment — serves one ultimate goal. Getting to the point where your standard isn't something you enforce, but something the team enforces on itself.

That happens when the people you've hired are good enough and cared-for enough to take ownership. When a crew member corrects a new hire's walkthrough because "that's not how we do it here" without you having to say anything. When the team genuinely doesn't want to let down the standard because they're proud of what the company is and what it stands for. When your reviews consistently reflect the experience you intended because every person on the job is committed to creating it.

You get there through hiring people with character, onboarding them into clear systems, developing their skills beyond the technical, recognizing their contributions, giving them a voice, and paying them fairly for their work. None of it is complicated. All of it is intentional.

The business that doesn't need you on every job isn't built through technology or systems alone. It's built through people who are capable, motivated, and invested in doing the work right. That's the team worth building.

> *A great team doesn't just get the job done. They make you look good when you're not watching.*

Go Deeper

For more on this and everything else in the book, the Haulers' Edge Newsletter goes deeper every week. Scan the code or subscribe free at HaulingHubb.com.

CH. 24 PAY LESS FOR EVERYTHING

"In business as in life, you don't get what you deserve, you get what you negotiate." — Chester L. Karrass

Every dollar you spend on vendors, suppliers, dump fees, fuel, equipment leases, and software subscriptions is a dollar that came off a job you worked to book, quote, and complete. Most of those costs feel fixed — the number on the invoice, the rate on the contract, the fee at the transfer station. Most of them aren't.

The price you're currently paying for the majority of your recurring business expenses is the price you accepted when you didn't ask for something better. That's not a criticism — it's just how vendor relationships work. Nobody calls you up and offers to charge you less. The responsibility to open that conversation is yours. And most operators never do.

A 10% reduction across your major recurring costs doesn't sound dramatic until you run the math. If your monthly overhead includes $3,000 in dump fees, $1,500 in fuel, $800 in equipment payments, and $700 in software and services, you're spending $72,000 a year on those four categories alone. Ten percent back is $7,200 — from conversations, not from working more jobs.

You Have More Leverage Than You Think

The common assumption is that negotiating is for big companies with purchasing departments and volume that matters. That assumption is wrong, and I've proved it personally.

A few years back our usual dump site closed for renovations

and we had to shift to a pricier private transfer station. We were a small operation — three trucks, a couple dozen dumpsters. Not exactly a high-volume account by any measure. I showed up in person and laid it out straight: we're a local company, we want to bring all our business here, and we need a rate that makes that sustainable for us.

We dropped from $125 per ton to $90. A 28% reduction just by showing up and asking. No threats, no leverage games, no negotiation tactics — just a direct conversation about what we needed and what we were willing to commit to in exchange.

That experience established something I've seen reinforced dozens of times since. Vendors — especially private operators, local businesses, and anyone who competes for accounts — want customers who show up consistently, pay reliably, and don't create problems. If you represent that kind of customer, you have more negotiating position than your volume alone would suggest. Loyalty, consistency, and long-term commitment are worth real money to suppliers, and they'll trade price concessions to secure them. Beyond the dump fee, the pattern holds across every vendor category. Shopping our commercial insurance to competing providers — and letting our current carrier know we were doing it — produced a savings of several thousand dollars a year when they negotiated rather than lose the account. Using the same mechanic for our entire fleet for years has bought us priority scheduling and transparent pricing that we'd never get as a new account. Running a fleet fuel card that also works at supply shops eliminated the need for drivers to circle back to the office mid-day — a time efficiency that compounds across every operating week. None of these required a confrontation. They required showing up consistently and being willing to have the conversation.

The worst outcome of any negotiation is the price stays the same. You're no worse off than before you asked. That asymmetry — nothing to lose, real money to gain — makes asking a straightforward decision.

Think Beyond the Unit Price

The mistake most operators make when they do negotiate is focusing only on the per-unit cost or the monthly rate. Price is one dimension of a deal. Terms are often where the real value lives.

A vendor who can't reduce their rate by 15% might be able to extend your payment terms from net-15 to net-45 — which meaningfully improves your cash flow without changing what you pay. A software company that won't discount the monthly fee might waive the setup cost, add user seats at no charge, or extend your contract at the current rate before a scheduled price increase. An equipment lessor might throw in a free maintenance period or reduce the early termination penalty in exchange for a longer commitment.

Before any vendor conversation, ask yourself: beyond price, what would make this arrangement better for my business? Cash flow timing. Service levels. Contract flexibility. Penalty reduction. Added services at current rates. All of these have real dollar value and are often easier to obtain than a straight price cut because they don't hit the vendor's revenue line as directly.

Build your negotiation around total value, not just the invoice number. When you frame the conversation that way — "I'm not necessarily asking you to cut your rate, I'm asking if there are ways we can structure this arrangement so it works better for both of us" — you open the door to creative solutions that a pure price negotiation would never surface.

Prepare Before You Pick Up the Phone

Walking into a negotiation unprepared produces worse outcomes and more anxiety than it needs to. The preparation is straightforward.

Know your numbers. Before you contact any vendor, calculate exactly what you're spending with them — per month, per year, per unit. How long have you been a customer? How consistently do you pay? What volume do you represent? These facts are your foundation. "We've been bringing you roughly 40 tons a month for the past two years — that's nearly $100,000 in business. We'd like to continue growing that, but we need to find a rate that makes it sustainable" is a different conversation than "can you do any better on price?"

Know the market. What are your alternatives? What do competitors charge for the same service? What would switching actually cost you in time, friction, and transition risk? You don't have to threaten to leave — but knowing your options gives you genuine confidence rather than performed confidence. Even mentioning that you've received competitive quotes changes the dynamic of the conversation.

Know your walk-away point. What's the minimum improvement that makes this conversation worthwhile? If the current dump rate makes a category of jobs unprofitable, what rate restores the margin? Define that number before the conversation so you're not making calculations under pressure.

Plan your opening. Lead with appreciation and history before you make the ask. The vendor relationship has value — acknowledge it. Then be direct about what you need and why. Most vendors respond better to clarity than to circling around an ask. "I'd like to talk about our rate structure. We've been a loyal customer and we want to keep this relationship long-term, but I need to understand if there's flexibility in the pricing." That's the whole setup. Everything else is a response to what they say.

The Conversation Itself

The tone that works best in vendor negotiations is collaborative rather than confrontational. You're not trying to squeeze them — you're trying to find an arrangement that works long-term for both sides. That framing is both strategically effective and genuinely true. A vendor who feels squeezed into a deal they resent will find ways to recover the margin — through service quality, through how they prioritize your orders during busy periods, through how quickly they respond when something goes wrong.

Be direct about your situation. Honesty in vendor negotiations is underrated. "These tipping fees are making certain job types unprofitable for us. We want to keep bringing you all our volume but we need to find a number that keeps that viable." No theatrics, no exaggeration, no manufactured urgency. Just the real situation explained plainly.

Ask open questions that invite solutions rather than yes-or-no responses. "Is there flexibility in pricing for customers who commit to a monthly volume?" "What would a longer-term agreement look like in terms of rate?" "Are there ways to structure this that would work better for both of us?" These questions keep the conversation moving toward creative options rather than a binary accept-or-reject dynamic.

When they make a concession — even a small one — acknowledge it specifically and appreciate it genuinely. That acknowledgment costs you nothing and reinforces the behavior you want to see in future conversations.

The Vendor Relationships Worth Building

Not every vendor relationship deserves the same investment. Focus your negotiation energy on the categories where the dollars are largest and the flexibility is greatest.

Dump fees and disposal costs are the first place to look in a hauling or junk removal operation. Transfer stations vary significantly in how they price — by weight, by volume, by material type, and by account status. Private operators typically have more pricing flexibility than municipal facilities. If you're using multiple facilities, consolidating volume at one location gives you a stronger negotiating position.

Fuel costs are partially negotiable through fleet fuel cards and bulk purchasing programs. Many fuel providers offer volume discounts, locked rates for specified periods, and rebate structures for consistent usage. If you're running multiple vehicles, a fleet fuel card program almost always saves money compared to paying at the pump on individual cards.

Equipment leases and financing are negotiable at origination — the terms on the initial agreement set the cost structure for the life of the lease. Most operators accept the first terms presented. Rates, residual values, maintenance inclusions, and early termination conditions are all negotiable. If you're renewing or adding equipment, get multiple quotes and use them against each other.

Software and service subscriptions feel fixed because they're presented as fixed. They're rarely actually fixed, especially for accounts that have been customers for a meaningful period. Annual billing versus monthly typically produces a 10–20% discount. Asking for a loyalty rate, a non-profit rate if applicable, or simply asking "is this the best available rate for our account" before renewal frequently produces a concession.

Insurance premiums are negotiable through your broker relationship, through shopping coverage at renewal, through bundling policies, and through demonstrating safety record improvements. A clean loss history over several years is leverage that rarely gets used.

Pay on Time, Every Time

The negotiating leverage that gets overlooked most consistently is your payment reliability. Vendors — especially smaller local operators — deal with customers who pay late, dispute invoices, and create accounts receivable headaches constantly. A customer who pays on time every time without being chased is genuinely valuable to them beyond the revenue.

When you establish that reputation — and it takes consistency over months, not a single payment — you can reference it directly in negotiations. "We've been on your books for two years and I don't think we've ever been late. That consistency has value and we'd like to see it reflected in our rate." That conversation works because it's true and because the vendor knows exactly how unusual reliable payment is.

Pay reliably, communicate proactively when something is going to be late, and never dispute an invoice without a legitimate reason. Those behaviors create a relationship foundation that makes every other negotiation easier.

Run the negotiations this chapter describes across your five largest vendor relationships. Dump fees, fuel, equipment, insurance, software. Not all at once — one conversation a month for five months. Most operators never have these conversations. That's why the money is still sitting there. That's why the money is still sitting there.

One more point for the operator who's already negotiated once and thinks there's nothing left to discuss: every major vendor relationship should be reviewed annually. Your business has grown. Your volume has changed. Market rates have shifted. The conversation you had two years ago is not the conversation you'd have today. A dump fee you negotiated when you were running two trucks is worth reopening now that you're running five — the volume leverage is different. A software subscription you signed at a promotional rate may have renewed at full price without you noticing. An insurance premium that felt fair when your fleet was smaller may have room to move now that you're a more established account. The operators who treat vendor negotiations as a one-time event leave money on the table every year. The ones who put it on the calendar keep finding it.

> *Every conversation you don't have is a cost you're choosing to keep. Pick up the phone.*

Go Deeper

For more on this and everything else in the book, the Haulers' Edge Newsletter goes deeper every week. Scan the code or subscribe free at HaulingHubb.com.

PART 5: EXPANSION & INNOVATION

CH. 25 STRATEGIC PARTNERSHIPS THAT PRINT MONEY

> *"No matter how brilliant your mind or strategy, if you're playing a solo game, you'll always lose out to a team." — Reid Hoffman*

Every customer you acquire through advertising costs you money. Google Ads, yard signs, door hangers, follow-up sequences — all of it has a cost that comes off the top of the job before you've made a dollar of profit. In our client base at Adimize, the average cost to acquire a new service business customer through paid search typically runs $50–$100 depending on market competitiveness, service type, and conversion rate — with well-optimized campaigns on the lower end and competitive urban markets on the higher. Your market may differ, but the range is a useful benchmark when evaluating whether a partnership referral that costs you nothing is worth more than it looks.

A referral from a trusted partner costs you nothing to generate.

The contractor who calls you every time a renovation project needs a dumpster isn't a lead you paid for. The real estate agent who refers every estate cleanout to you isn't coming from your Google Ads budget. The moving company that sends you every customer who has leftover junk after the truck is loaded didn't cost you a single dollar in customer acquisition. These are partnerships — and they're one of the highest-return growth strategies available to a local service business.

Most operators know this in theory. Few build it systematically. This chapter is about building it.

Why Partnerships Beat Most Marketing

The math is simple. If acquiring a customer through paid channels costs $75 and your average job generates $150 in gross profit, you're keeping half. If a partnership referral generates that same job at zero acquisition cost, you keep the whole margin.

That's not the only advantage. A referred customer arrives pre-sold. They're not comparing three quotes — they called you specifically because someone they trust told them to. That increases your close rate, reduces the friction in the booking conversation, and often produces a better customer relationship because the trust transferred from the referring partner.

Referred customers also tend to refer others. Someone who found you through a trusted recommendation and had a great experience is more likely to become a referral source themselves than someone who found you through an ad. The compounding effect of a strong referral network grows over time in a way that paid acquisition never does — because paid acquisition stops producing the moment you stop paying.

According to Nielsen, 92% of consumers trust referrals from people they know more than any other form of advertising *(Nielsen Global Trust in Advertising Report, 2021)*. That trust differential is the fundamental advantage of partnership-based business development over paid marketing.

The Partnership Categories That Produce Real Volume

Not all partnerships are equal. The ones worth building are those where your services and your partner's services naturally touch the same customer at the same moment of need — where the referral feels genuinely helpful rather than forced.

Real estate agents and property managers. These are the highest-volume referral sources for most hauling and junk removal operations. Real estate agents list homes constantly and those homes consistently need cleanouts before listing — garages, basements, attics, hoarding situations. Property managers turn over units on a regular cycle and every turnover potentially involves junk removal, cleanout, or a dumpster for renovation debris. One active real estate agent or property management company can generate dozens of jobs a year from a single relationship. These are relationships worth pursuing aggressively and maintaining deliberately.

Contractors and renovation companies. A contractor who trusts you becomes a recurring revenue stream. Every kitchen gut, every bathroom demo, every deck tear-off generates debris that needs to go somewhere. If you're the company that delivers a dumpster the day the project starts, swaps it when it fills, and picks it up the day work wraps — without drama or delay — you become part of their workflow. Getting removed from their workflow requires an active decision on their part. You become the default because switching has friction.

Moving companies. Movers constantly encounter customers who have items they're not taking to the new place — furniture, appliances, accumulated junk that needs to go. The moving company doesn't handle it. You do. A simple referral arrangement with a local moving company — they mention you to every customer who has leftover items — produces consistent inbound calls from pre-qualified prospects who are motivated to act quickly because they're in the middle of a move.

Adjacent service trades. Cleaners, landscapers, HVAC companies, painters, pest control operators — all of these businesses work in the same homes and properties you work in. Their customers regularly need services adjacent to yours. An estate cleaner working on a property that needs junk hauled doesn't do junk hauling. A landscaper dealing with a yard full of debris doesn't have a dump truck. These natural overflow moments are referral opportunities that both parties benefit from formalizing.

Restoration and remediation companies. Water damage, fire cleanup, mold remediation — all of these situations generate significant debris removal needs. Restoration companies work on tight timelines and need reliable vendors they can call at any hour. Being the junk removal company that a restoration contractor can count on for fast response positions you for some of the highest-margin, most urgent work in the market.

Competitors with different specializations. This one surprises people but it works. A junk removal company that doesn't do demolition needs someone to refer demo work to. A demo contractor who doesn't rent dumpsters needs a dumpster partner. A company that doesn't serve your geography needs someone who does for overflow. These relationships are worth building carefully — with clear boundaries around what's being referred and what isn't — because they convert overflow and out-of-scope requests into revenue rather than lost opportunities.

Building a Partnership That Lasts

The first conversation with a potential partner isn't a sales pitch. It's an exploration of whether there's genuine overlap that serves both businesses. The question isn't "will you send me business" — it's "do our customers regularly need both of our services, and is there a way to make sure they find both of us?"

That framing changes the dynamic. You're not asking for a favor. You're proposing a mutually beneficial arrangement that makes both businesses more useful to their customers.

Make contact where business owners naturally gather — Chamber of Commerce events, trade association meetings, local business groups. LinkedIn works for direct outreach, especially to property managers and real estate agents who use it actively. But understand what you're walking into. LinkedIn inboxes in 2026 are saturated with automated connection requests and templated messages that all read the same way. A generic text message asking to connect gets lost in the noise — personalized messages consistently outperform generic ones by a significant margin — in our own outreach we've seen response rates two to three times higher on personalized video messages versus templated text, and that gap keeps widening as automation makes generic outreach cheaper and more common.

The way to break through is effort that's visible. Instead of sending a text message, send a voice note or better yet a short video — thirty seconds, their name said out loud, something specific about their business that signals you actually looked them up. Video outreach on LinkedIn consistently generates dramatically more responses than text — LinkedIn's own platform data shows video posts generating significantly higher engagement than text, and in direct outreach the difference is even more pronounced. The same principle applies in a direct message. A video that opens with "Hey Sarah, I noticed you manage properties in the Hartford area — I run a junk removal and dumpster company serving the same market and I think our customers overlap. Would love thirty seconds of your time" gets opened. A text message that says the same thing mostly doesn't.

It takes longer. You'll send fewer messages in an hour than someone blasting a template. That's the point. Laser-focused outreach to twenty well-researched contacts outperforms a generic blast to two hundred every time. The contact who receives a personalized video from someone who clearly did their homework is already predisposed to say yes before they've heard a single word about what you do.

Start with a small test before committing to anything formal. Refer a customer to them and see how they handle it. Take a referral from them and make sure you deliver at the level that reflects well on them for sending you. The first few exchanges establish whether the partnership is worth formalizing. A partner who disappoints your referral isn't just a service quality problem — it's a reputation problem for you, because you made the recommendation.

Once you've established that the quality is there on both sides, put the basics in writing. Not a complex legal agreement — a simple document that captures the referral structure, how you'll track jobs, how payment or reciprocity works if applicable, and how you'll handle a situation where a referral goes poorly. This doesn't need to be adversarial. It just needs to be clear. The partnerships that break down usually do so because expectations were never explicit.

Check in regularly. A quarterly conversation — even just a fifteen-minute coffee — keeps the relationship alive and gives both parties a chance to flag what's working, what isn't, and where new opportunities might exist. Partnerships that run on autopilot without maintenance tend to drift and eventually go quiet.

The Metrics That Tell You If It's Working

A partnership that feels productive but isn't tracked is just a pleasant relationship. Track every referral the same way you track every other lead source.

Log where every new job came from. If you're using a CRM, tag the lead source on every record — this is what builds the dataset for the 80/20 analysis from Chapter 12. Over time you can see exactly which partner relationships are generating volume and revenue and which ones exist mostly in theory.

Track both directions. How many referrals are you sending each partner and how many are you receiving? A relationship that's consistently imbalanced — you're sending ten jobs for every two you receive — needs a direct conversation before resentment builds. Most of the time the imbalance isn't intentional. The partner isn't aware of it because they're not tracking it either. The conversation — "I've sent you fourteen jobs this quarter and I've received three — can we talk about how to improve that flow?" — is almost always received well when it's framed as an observation rather than an accusation.

Track the revenue quality of partnership referrals separately from paid lead sources. Are referred customers booking at higher rates? Are they higher average ticket jobs? Are they generating better reviews? That data tells you the real value of the partnership network relative to your paid acquisition spend and informs how much time and attention to invest in building it.

If a partnership isn't producing after six months of genuine effort on both sides, have an honest conversation about why and whether it's worth continuing. Some partnerships look logical on paper but don't produce in practice — different customer demographics, different service areas, mismatched volume levels. Exit those cleanly and redirect the energy toward relationships with higher potential.

Your Partnership Network as a Competitive Moat

Here's the long-term picture that most owners don't consider when they think about partnerships.

A competitor can match your pricing. They can run Google Ads in your market. They can wrap their trucks and knock doors in your neighborhoods. What they can't replicate quickly is a relationship network you've spent years building — ten contractors who call you first, five property managers who have your number saved, three moving companies who mention you to every customer.

That network is a competitive moat. It took time to build, it requires consistent maintenance to keep, and it generates revenue at near-zero acquisition cost. A new entrant to your market — even a well-funded one — starts with none of it. They have to earn every one of those relationships from scratch, and the relationships you've already built create switching costs for the partners who value you.

The operators who build this network deliberately, maintain it consistently, and measure it honestly end up with a business development engine that produces at a fraction of the cost of paid marketing — and becomes more valuable every year as the relationships deepen and the referral volume compounds.

Start with three potential partners this month. One real estate agent or property manager. One contractor in a complementary trade. One adjacent service business whose customers regularly need what you do. Have the first conversation. Make the first referral. Begin building the network one relationship at a time.

> *The best lead you'll ever get is one a trusted partner sends you. Build the network that makes that happen consistently.*

Go Deeper

For more on this and everything else in the book, the Haulers' Edge Newsletter goes deeper every week. Scan the code or subscribe free at HaulingHubb.com.

CH. 26 HIDDEN REVENUE STREAMS

"Opportunity is missed by most people because it is dressed in overalls and looks like work." — Thomas Edison

Every job you run generates more value than what's on the invoice. The furniture that gets donated instead of dumped saves you a tipping fee. The scrap metal in the back of the truck has a cash value at the yard. The baseball cards in the bottom of a box that looked like junk turned out to be worth several hundred dollars. The resale platform on your phone can move a barely-used lawn mower before you get back to the yard.

None of these are your core business. All of them are money your core business is already generating that most businesses leave on the table.

This chapter covers three primary revenue streams that exist inside the work you're already doing — donations that save disposal costs and build your brand, scrap metal that converts waste into cash, and reselling items that others have paid you to remove. It also touches on a few additional opportunities worth knowing about depending on your trade and market. None of them should ever take priority over your paying jobs. All of them are worth building into your operation as systems rather than afterthoughts.

Donations: The Revenue Stream That Saves Money

Donating items instead of dumping them isn't charity — it's a margin improvement strategy with a community benefit attached.

Every pound of material you divert from the landfill is weight you don't pay to dump. If your transfer station charges $70 per ton and your crew diverts a ton of usable furniture in a month, that's $70 back in your pocket before you've sold a single item. Over a year, a consistent donation program on a busy operation can save thousands in tipping fees while simultaneously producing the kind of community goodwill that generates referrals.

The principle applies across service trades. For a landscaper it's green waste and yard debris diverted from disposal fees. For a cleaning company it's furniture and household goods left behind after an estate job. For a mover it's the items clients decide at the last minute aren't coming to the new place. The material differs by trade — the cost savings and community benefit are the same.

The customer-facing benefit is real too. When a homeowner asks where their stuff is going and you tell them usable items go to local shelters and donation centers while everything else is disposed of responsibly, that answer matters to a meaningful segment of your market. It differentiates you from every competitor who says they dump everything. And it gives your crews something to feel good about on the job — which has its own retention and morale value.

The challenge is execution. Donation programs that run haphazardly add cost rather than saving it. Extra trips to donation centers, time spent sorting items that turn out to be unacceptable, crews handling fragile items with more care than efficiency requires — all of these erode the margin benefit if they're not managed with discipline.

The system that works is built around three rules. Touch items as few times as possible. Plan donation drop-offs into existing routes rather than making separate trips. And know what each local center accepts before you show up.

Train your crew on a clear standard for what's donatable — clean, functional, safe. Solid furniture, clothing in good condition, working small appliances, books, toys. Not mattresses with stains, broken lamps, safety-recalled baby gear, or anything moldy. A one-page cheat sheet in every truck eliminates the guesswork and ensures consistent decisions in the field.

Organize the truck with donations in mind. Many operators dedicate a section of the truck to donatable items so they're loaded separately and don't get buried under debris. Color-coded tags work well — green for donate, blue for recycle, red for trash. Simple systems save minutes on every job that compound into hours over a week.

Build relationships with your local donation centers. Know their hours, their acceptance policies, and their staff by name. A center that knows your company will often make the process smoother — faster intake, fewer rejections, and occasionally referrals when community members ask where to find a reliable hauler. Call ahead before arriving with bulky items — "Are you taking sofas today?" — to avoid a wasted trip.

Track what you divert. Rough numbers are enough — weight or item counts, estimated tipping fee savings, donation center visits per month. The data builds accountability internally and gives you something concrete to mention when marketing your eco-friendly approach to customers. It also tells you whether the program is actually saving money or costing you time that exceeds the disposal savings.

The honest framing for customers is: we donate what we can to local organizations and recycle what's recyclable. We dispose of the rest responsibly. Don't overpromise — not everything is donatable and your partners don't want to be dumping grounds for broken junk. But that honest statement, delivered with the evidence to back it up, is a genuine differentiator.

Scrap Metal: Converting Waste Into Cash

Metal is everywhere in a service operation and most of it has value at a scrap yard. This isn't limited to hauling. HVAC companies pull copper and aluminum on every equipment swap. Plumbers handle brass and copper fittings constantly. Contractors generate significant metal debris on every renovation. Cleaners and property managers encounter appliances and metal fixtures regularly. If your trade touches metal — and most do — there's a scrap revenue stream sitting in your existing workflow. Old appliances, pipes, bed frames, radiators, construction debris, HVAC components — all of it can be converted to cash rather than paid to dump.

The foundation is a dedicated scrap zone at your facility. A single bin or container specifically for metal, clearly designated and consistently maintained. When your crew knows exactly where metal goes, it actually goes there. Without the designated zone, metal gets mixed into general waste and the value disappears into the landfill.

Separate ferrous from non-ferrous at even the most basic level. Ferrous metals — steel, iron, anything a magnet sticks to — pay less per pound but make up most of your volume. Non-ferrous metals — copper, brass, aluminum — pay significantly more and are worth pulling aside into a dedicated bin. A pound of copper can fetch several dollars while a pound of steel might bring pennies. The math on even modest volumes of copper pipe, brass fittings, and aluminum makes the separation worth a few minutes of sorting at the yard.

For the bulk ferrous scrap — appliances, steel frames, mixed metal debris — keep it simple and let the yard do the fine sorting. They have magnets and machines designed for it. The temptation is to strip every copper wire and dismantle every appliance in the field. Unless you have genuine downtime and a crew member who enjoys the work, that level of micro-sorting consumes more labor than it returns. Your crew's time on a paid job is worth far more than the incremental scrap value of a stripped motor.

Batch your scrap runs. The single biggest efficiency mistake in scrap operations is making too many small trips. A half-load trip to the yard costs you the same fuel and labor as a full load but generates a fraction of the revenue. Let the bin fill. Schedule scrap runs during slow periods — early mornings, between jobs, end of the week. One well-timed full load beats four partial ones by a wide margin.

Some operators tie scrap returns to team motivation — a monthly crew lunch funded by the scrap check, or a contribution to a tool fund. That buy-in matters. When crew members see a tangible benefit from taking the extra thirty seconds to throw metal in the right bin, they do it consistently. Without it, the habit erodes.

Track the numbers. A simple log — date, weight, payout — takes five minutes per run and tells you whether the program is generating meaningful money or mostly keeping you busy. In most active hauling operations, consistent scrap collection produces several hundred to over a thousand dollars a month depending on volume and material mix. That range is wide enough to justify tracking rather than estimating.

Reselling: The Items Worth More Than the Dump Fee

Items your customers have paid you to remove occasionally have real resale value. This opportunity is most visible in junk removal where the volume and variety of material is highest, but it exists in any service business that handles customer property. Estate cleaners, movers, property managers clearing units, and restoration companies all encounter items with resale value in the course of normal operations. The discipline of recognizing and capturing that value is the same regardless of trade. That value is legitimate revenue — you were paid to take the item, it's yours to dispose of responsibly, and selling it rather than dumping it is simply the most profitable form of responsible disposal.

The key word is occasionally. Most junk is junk. But in a business that moves the volume of material a busy service operation does, the exceptions add up. An old wooden dresser in solid condition. A working lawn mower. A set of baseball cards with autographed Willie Mays tucked in a binder. Original artwork that looks like prints until you look at the signature. These finds happen. The operator who's trained to recognize them captures the value. The one who isn't loses it to the landfill.

Train your crew on a simple filter: would someone buy this? If the answer is yes and it's in condition that doesn't require significant repair time, set it aside. If it needs more work than the likely resale value justifies, donate it or dump it and move on. Don't create a death pile of items you intend to sell eventually. Eventually doesn't happen in a busy operation.

Facebook Marketplace is the right platform for 80–90% of what you'll want to move. Free, local, fast. Post a photo, a clear description, a fair price based on what comparable items are selling for in your area, and it's usually gone within a week. OfferUp and Craigslist work for larger items and tools. eBay makes sense for collectibles, specialty items, or anything with national demand that justifies the shipping logistics. For most day-to-day resale in a service business context, Facebook Marketplace is where the buyers are.

The listing that sells quickly has four elements: clear photos in decent light, an honest description including any flaws, a searchable title with relevant keywords rather than a vague label, and a price slightly below comparable listings so it moves fast. You acquired the item for free or were paid to take it — you can afford to price competitively. "Solid wood 6-drawer dresser, some minor scratches on top, otherwise excellent condition — $75" moves faster than "Dresser, good condition — $100."

Early in the business we started offering to take boxes of books customers were clearing out. We'd sort them — donations for most, Amazon FBA for anything that looked valuable. The FBA model worked — boxes shipped to their warehouse, they handled the rest, monthly checks arrived. But once we did the honest math on cataloging, boxing, and shipping time, and then asked what it would cost to pay someone else to run that operation, the margins evaporated. A side hustle that makes sense when you're doing all the labor yourself often doesn't survive the transition to a managed operation.

That lesson applies broadly to resale. It works as a bonus layer on an existing operation when it's disciplined, time-bounded, and selective. It fails when it becomes its own operation competing with your core business for attention and resources. The $500 client job always comes before the $50 Marketplace flip. Build time boundaries — specific windows for listing, responding, and coordinating pickups — and hold them.

Manage resale communication with the same professionalism as your client communication. Use a dedicated account or phone number so buyer messages don't tangle with customer calls. Respond promptly but set clear expectations. Communicate your pickup logistics clearly. Handle cash carefully and note the "as-is" nature of every sale in the listing. Resale reputation in your local market compounds the same way your service reputation does — do it professionally and buyers come back, refer others, and leave reviews on your selling profile.

These three get full treatment because they're the most universally applicable. But depending on your trade and market, other streams are worth exploring: upselling adjacent services to customers mid-job (pressure washing a driveway while doing a junk removal, for example), renting equipment during downtime, or running loads for other small operators who don't have your disposal relationships. None of these are core revenue strategies. All of them are worth knowing exist.

Keep It in Perspective

These three revenue streams share a common principle: they extract additional value from work you're already doing without adding material cost to the operation. Donations reduce disposal fees. Scrap converts waste into cash. Resale turns discarded items into found money.

None of them should distort your priorities. Your paying jobs, your customer experience, your marketing, and your team come first. These are bonus loops, not core revenue strategies. But treated as systems — built into your SOPs, trained into your crew, tracked in your numbers — they compound into meaningful annual revenue that improves your margin without requiring additional sales effort.

The operator who runs a consistent donation program, maintains a disciplined scrap system, and moves resaleable items efficiently through Marketplace is generating real money from material that competitors are paying to bury. That's not a side hustle. That's smart operations.

> *Every load you run has more value in it than you're currently capturing. Build the systems to capture it.*

Go Deeper

For more on this and everything else in the book, the Haulers'

Edge Newsletter goes deeper every week. Scan the code or subscribe free at HaulingHubb.com.

CH. 27 THE RECURRING REVENUE PLAYBOOK

> *"Someone is sitting in the shade today because someone planted a tree a long time ago." — Warren Buffett*

The most stressful version of a service business is one where every week starts at zero. No guaranteed work. No predictable revenue. Just whatever you can book before Friday.

The most stable version is one where a meaningful portion of your schedule is already filled before the week starts — contractors who call every month, property managers who have you on their vendor list, commercial clients whose ongoing needs generate steady work without you running a new sales process every time.

The difference between those two versions is recurring revenue. And building it is less about selling and making yourself the obvious, indispensable choice, and more about finding the right clients to pursue.

One of my first real recurring relationships came through a local real estate investor. Over a single summer that one relationship turned into well over six figures in annual recurring revenue. No selling after the first job. Just showing up, delivering, and the next call coming in. That experience made something clear that I've operated from ever since: a handful of steady clients who trust you is worth more in revenue and peace of mind than dozens of one-time jobs from strangers you have to re-acquire every week.

Who Has Recurring Needs

Not every customer is a recurring revenue opportunity. The ones worth pursuing are the ones whose responsibilities naturally create repeat demand — where the need isn't one event but an ongoing condition of how they operate.

Contractors and trade professionals sit at the top of this list for most service businesses. Builders, remodelers, landscapers, electricians, plumbers — almost every project they touch generates ongoing needs. Their work has natural rhythms and repeat tasks. A contractor who's always demo-ing something always needs debris removed. A landscaper who's always finishing jobs always needs haul-away. If you position yourself as the partner who removes friction from their workflow rather than just a vendor they call occasionally, you become part of their system. That's a fundamentally different relationship than a one-off customer.

Property managers and facility operators are another primary category. They're responsible for keeping properties occupied, maintained, and turning over efficiently — which means cleanouts, unit prep, and maintenance support are never-ending. Every tenant turnover is an opportunity. Every renovation cycle is an opportunity. Once you're on their preferred vendor list and they know they can hand you a key and walk away without worrying, you're not competing for that business anymore — you're just receiving it.

Beyond those two core categories, recurring opportunities exist with real estate investors and agents who prep properties repeatedly, storage facilities dealing with abandoned units on a regular cycle, banks and asset managers handling foreclosures, restoration and remediation companies that generate debris on every project, and any business with predictable seasonal needs. The specific category matters less than the underlying principle: you're looking for people whose job description creates ongoing demand for what you do.

For a landscaper, the recurring client is a property management company that needs grounds maintenance on a regular schedule. For a cleaning company, it's the commercial account with weekly service. For an HVAC company, it's the service agreement customer who calls for seasonal maintenance. The specific service differs by trade. The strategic value of recurring revenue is identical across all of them.

Making the Approach

Recurring clients rarely come to you on their own. You have to make the move — and the move needs to be positioned around their problems, not your services.

Contractors are busy, often overwhelmed, and focused on the next deadline. The pitch that lands isn't about what you do — it's about what you remove from their plate. "I know you're managing multiple job sites and debris removal is probably the last thing you want to think about. If you need same-day turnaround or after-hours service, that's what we do. I'd like to be the company you call when that problem comes up." That framing positions you as a solution to a real operational pain rather than another vendor asking for business.

Catch them where they are — on a job site, at a trade association meeting, through a direct LinkedIn message, or a straightforward phone call. Don't overcomplicate the outreach. A brief, confident introduction that demonstrates you understand their world gets a response more often than operators expect.

Property managers want vendors who reduce friction. Their days are full of maintenance crises, demanding landlords, and tenant issues. The last thing they need is a service provider who creates more work. The pitch is simple: call me once and it's handled. Reliability, insurance, professionalism, and fast turnaround. If they know they can trust you with a key and a brief description of the job, you've differentiated yourself from 90% of the vendors on their list who require hand-holding.

A brief introduction — a short email, a drop-in with a card, a phone call — gets the conversation started. What seals the deal is the first job. Treat every first job with a potential recurring client as an audition. Tell your crew explicitly. If that job goes smoothly, there will be more of them. One property manager I worked with handed us a trashed rental unit early on. We cleared it fast and left it broom-swept. She called back the next week: "Can we just make you our standing cleanup company?" That single yes turned into years of steady work.

Let the Relationship Define the Structure

Most recurring service relationships don't need a formal written agreement. If a contractor calls you every time a project generates debris because you've earned that position, introducing a one-page contract into that dynamic adds friction to something that's already working. The agreement that matters in those relationships is the one you make every time you show up and do the job right.

Where written clarity does matter is when the volume and complexity justify it. A property management company responsible for fifty units, a commercial client with multiple locations, or any account large enough that a pricing misunderstanding would be a significant problem — these relationships benefit from documented terms. Not a legal contract, but a simple written confirmation of scope, pricing structure, billing terms, and how either party can end the arrangement if needed. One page. Plain language. Enough to eliminate the misunderstandings that derail high-value relationships when expectations were never explicit.

The cancellation provision is worth including when you do

formalize things — not as a trap but as a comfort. A client who knows they can exit with thirty days' notice is more willing to commit than one who feels locked in. That clause almost never gets used. If you keep delivering they won't want to leave. It's there to make saying yes feel safe.

Billing convenience matters regardless of whether anything is in writing. A contractor managing five active projects doesn't want to approve a separate invoice for every job. Consolidated monthly billing is an immediate differentiator. Net-30 terms for property managers fit their accounts payable cycle. These accommodations move you from the annoying vendor category to the easy to work with category — and easy vendors get the repeat calls while complicated ones get replaced.

The best recurring relationships run on performance, not paperwork. Earn the call. Keep earning it. For the accounts where the stakes are high enough to document, keep it simple. For the ones built entirely on trust and track record — let the work speak and stay out of its way.

One note for the operator thinking long-term: if you ever plan to sell the business, informal recurring relationships will need to be documented. A buyer's attorney will ask for evidence of those revenue streams. A handshake arrangement has no value in a transaction — a written service agreement does. Chapter 29 covers this in detail. You don't need paperwork to run the relationship well. You'll want it if you ever want to sell what you've built.

Make Yourself Indispensable

Landing the recurring relationship is step one. Keeping it long enough for it to compound is step two. That happens through one mechanism: making yourself harder to replace than the cost savings from switching would justify.

The contractors and property managers who have stuck with

me longest aren't staying because of price. They're staying because switching would require them to find someone new, explain their systems, go through the reliability test again, and accept some level of service uncertainty during the transition. That switching cost is real and it grows with every month of consistent, reliable performance.

The specific behaviors that build that switching cost are simple. Show up when you say you will, every time. Respond to messages within the hour during business hours. Deliver the one or two extra touches that your client actually values — for contractors it might be same-day turnaround on urgent calls, for property managers it might be broom-sweeping after every cleanout, for a commercial client it might be before-and-after photos sent immediately after each service. One contractor told me early on that if my driver could call 30 minutes before arriving it would help them prep the site. Easy adjustment. It locked us in tighter than any discount would have.

Priority scheduling for recurring clients is another mechanism worth building explicitly into your operation. Clients who know they're at the front of your schedule during peak season — that you've blocked time specifically for them during your busiest periods — feel valued in a way that creates loyalty money can't replicate. One of my best property manager relationships was built partly on the fact that every last week of the month, prime move-out season, we had her units blocked on the schedule. She never had to stress about availability. We always knew where part of our week was going.

Ask your recurring clients regularly what you could do better. Not once at the start and then never again — periodically, genuinely. A client who feels heard and sees their feedback implemented becomes more invested in the relationship than one who's just satisfied with the service.

Grow the Relationship Over Time

A contractor who trusts you for one service often needs several. A property manager who relies on you for unit cleanouts may also manage commercial properties. The clients you already have are your best source of expanded business — but only if you ask.

"Is there anything else we can help with?" is a sentence that rarely gets said to a business's best clients. Ask it. Even when the answer is no, you've signaled that you're interested in serving more of their needs, which strengthens the relationship. When the answer is yes, you've grown the account without spending anything on acquisition.

Ask for referrals deliberately. Contractors know other contractors. Property managers talk to other managers. A simple "Do you know anyone else who could use what we do for you?" is one of the highest-return questions in a service business. A warm referral from a trusted recurring client converts at a rate that no paid marketing channel can match.

Ask for testimonials when the relationship is strong. A line from a respected property manager in your market — "We rely on them for all our rental cleanouts and they've never let us down" — carries more weight in a prospective client's decision than any ad you'll run. Your recurring clients are your best marketing asset. Most operators never activate them.

The Compounding Math

Here's why this matters enough to build a strategy around.

A one-time residential customer who books a $400 cleanout generates $400. Acquiring them cost you $75 in marketing. Net contribution before operating costs: $325.

A contractor who generates two jobs a month at $400 each generates $9,600 per year. Acquiring them cost you one introduction and one audition job. Net contribution before operating costs: substantially more per dollar of acquisition spend than you'll get from any paid channel.

A property manager responsible for fifty units who calls you for every turnover — even at a conservative eight turnovers a year at $350 each — generates $2,800 annually from a single relationship you spent an afternoon building.

This isn't about working less. It's about working on the right things. Every hour you invest in building and maintaining a recurring client relationship has a higher expected return than the same hour spent generating one-time leads. The revenue is more predictable, the acquisition cost is lower, the customer experience is better because you know each other, and the relationship itself becomes a referral source for more relationships like it.

Recurring revenue is the difference between a business that feels like a treadmill and one that feels like it's building toward something. Build the base.

> *The clients worth having are the ones who call you without shopping around. Earn that position and protect it.*

Go Deeper

For more on this and everything else in the book, the Haulers' Edge Newsletter goes deeper every week. Scan the code or subscribe free at HaulingHubb.com.

CH. 28 ONE ACCIDENT CAN END EVERYTHING

> *"Risk comes from not knowing what you're doing." — Warren Buffett*

You've built something real. Clients who trust you. A crew that shows up. Revenue that's growing. Systems that are starting to run without you needing to be everywhere at once.

One accident — one serious injury, one property damage claim, one lawsuit — can take all of it. Not slow it down. Take it. The settlement that exceeds your coverage. The judgment that pierces your business entity and reaches your personal assets. The workers' comp claim that exposes you because you classified someone incorrectly. The vehicle accident where your driver was at fault and your commercial auto policy lapsed because you missed a payment.

None of these are hypotheticals. They happen to real operators in real service businesses every year. The ones who survive them had the right coverage and the right structure in place before they needed it. The ones who don't survive often didn't know what they were missing until the claim arrived.

This chapter is about understanding the specific exposures your business carries and closing them before someone else does it for you.

The Entity Structure That Protects Everything You Own

We covered LLC and S-Corp elections in Chapter 6 from a tax perspective. The other reason entity structure matters is liability protection — and for a service business operating vehicles, entering customers' properties, and employing people, that protection is non-negotiable.

A sole proprietorship — which is what you are if you haven't formed a legal entity — provides zero separation between your business and your personal life. If your business gets sued and loses, the judgment can reach your personal bank account, your home, your savings, everything you own. There is no legal wall between what you've built professionally and what you've built personally.

An LLC creates that wall. Not an impenetrable one — courts can pierce the corporate veil in cases of fraud, commingling of personal and business funds, or failure to maintain the entity properly — but a real and meaningful one when maintained correctly. The business gets sued, not you personally. Your personal assets are protected as long as you're operating the business as a genuine separate entity.

Maintaining the protection requires following a few basic rules consistently. Keep business and personal finances completely separate — separate bank accounts, separate credit cards, no personal expenses run through the business and no business expenses run through personal accounts. Pay yourself formally rather than just pulling cash from the business account. Keep basic records of major business decisions. These requirements are not onerous. Failing to follow them is what exposes you.

If you're currently operating as a sole proprietor, forming an LLC is a few hundred dollars and an afternoon of paperwork in most states. It's the most immediate risk reduction action available to a service business owner who hasn't done it yet. Do it before you need it.

General Liability Insurance:

Your First Line of Defense

General liability insurance covers bodily injury and property damage claims arising from your business operations. For a service business — where you're in customers' homes and on job sites every day — this is the foundational coverage that makes everything else possible.

If a crew member drops something and damages a customer's floor, general liability pays. If a customer trips over your equipment and gets injured, general liability pays. If your work causes property damage that results in a claim, general liability pays. Without it, those claims come directly out of your business — and potentially through your business into your personal assets.

The minimum coverage most service businesses should carry is $1 million per occurrence and $2 million aggregate. For businesses working with commercial clients, property managers, or contractors, $1 million is often the minimum they'll require before putting you on their vendor list. Some require $2 million per occurrence. Know what your clients require and make sure your policy meets it.

The cost is more manageable than you might expect — typically $1,000 to $3,000 annually for a small service operation depending on your revenue, number of employees, and the nature of the work. Compare that to the cost of a single uninsured claim. The math is not close.

Work with an independent insurance broker who specializes in small businesses or specifically in service and contracting trades. They can compare coverage across multiple carriers, identify gaps in standard policies, and ensure your coverage matches your actual operations. A retail insurance agent who primarily sells auto and home policies is not the right person to structure commercial liability coverage for a service business.

Commercial Auto: The Exposure Most Operators Underestimate

Your personal auto insurance policy does not cover your vehicle when it's being used for business purposes. This is one of the most common and most expensive gaps in service business insurance.

If you're driving a personal vehicle to and from job sites, using it to haul equipment, or operating it in any way connected to your business — and your policy is personal auto — you are uninsured for business use. If you have an accident while on a job and your personal insurer discovers the vehicle was being used commercially, they can deny the claim entirely.

Commercial auto insurance covers vehicles used for business purposes. It's priced differently from personal auto — typically higher, because commercial vehicles are on the road more hours per day and carry more liability exposure — but it's the coverage that actually applies when a business vehicle is involved in an accident.

If you're running a fleet of trucks, every vehicle needs commercial auto coverage. If you're using a personal vehicle for business purposes, you need either a commercial policy or a specific business use endorsement added to your personal policy. Talk to your broker about your actual vehicle usage and make sure the coverage matches.

The accident that happens while you're driving between jobs — in a truck with your company name on the side — is a business accident. Make sure your coverage treats it that way.

Workers' Compensation: Required and Non-Negotiable

In most states, workers' compensation insurance is legally required the moment you have your first employee. In a physical service business — lifting, loading, operating heavy equipment, working in varying environmental conditions — the injury exposure is real and consistent.

Workers' comp covers medical expenses and lost wages for employees injured on the job. Without it, an injured employee can sue your business directly for those costs, and in many states you can be personally liable for the full amount of the claim plus penalties for failing to carry required coverage.

The cost of workers' comp is calculated as a percentage of your payroll, adjusted for the risk classification of the work your employees do. For physical service trades, rates are higher than for office work — but the alternative to paying those premiums is self-insuring against injuries that can easily reach six figures in medical and lost wage costs.

Get workers' comp before your first employee starts, not after. The paperwork takes days. An injury can happen on day one.

If you're working with subcontractors, verify that they carry their own workers' comp coverage and get certificates of insurance before they step foot on a job. A subcontractor without their own workers' comp who gets injured on your job site may be reclassified as your employee for workers' comp purposes, leaving you responsible for the claim. Request certificates of insurance from every subcontractor you use and keep them on file.

The Certificates of Insurance That Protect Your Business Relationships

A certificate of insurance — COI — is a document your insurance broker generates that proves your coverage is in force. Every commercial client, property manager, contractor, and facility operator you work with should have a current COI on file for your business.

This is standard practice in commercial service relationships and most professional clients will request it before the first job. Have your broker generate COIs quickly when requested — delays in producing insurance documentation create delays in getting on vendor lists and delays in getting paid.

The reverse is equally important. When you hire subcontractors or enter into partnerships, request their COIs before any work begins. If something goes wrong on a job involving an uninsured subcontractor, your business may carry the exposure. A few minutes of paperwork before each new subcontractor relationship begins is inexpensive risk management.

Keep a file — physical or digital — with current COIs for every subcontractor you use regularly. Review them at renewal time to confirm coverage hasn't lapsed. An expired COI is the same as no COI from a liability perspective.

Property and Equipment Coverage

Your trucks, trailers, dumpsters, and equipment represent significant capital investment. A commercial property policy or inland marine policy — which covers equipment that moves from location to location — protects that investment against theft, damage, and loss.

Most general liability policies do not cover your own equipment. The truck that gets damaged in an accident needs commercial auto coverage. The dumpster that gets stolen from a job site needs a separate property or inland marine policy. The equipment stored at your facility needs commercial property coverage.

Review your current coverage and identify what's insured and what isn't. The gaps are often surprising. A piece of equipment worth $20,000 that isn't covered by any policy is a $20,000 exposure you're carrying unknowingly.

The Umbrella Policy: When Standard Coverage Isn't Enough

A commercial umbrella policy provides excess coverage above the limits of your underlying policies — general liability, commercial auto, and workers' comp. It kicks in when a claim exceeds the limits of your primary coverage.

This matters because serious claims don't always stay within standard policy limits. A significant injury, a multi-vehicle accident, a major property damage claim — these can generate liability that exceeds a $1 million general liability policy. An umbrella policy — typically $1 million to $5 million in additional coverage — is relatively inexpensive compared to the base policies it sits on top of, because it only pays after the underlying policies are exhausted.

For a service business with employees, vehicles on the road daily, and regular access to customers' properties, a $1 million commercial umbrella policy is worth serious consideration. Talk to your broker about whether your current exposure justifies it.

Employment Practices Liability

As your team grows, a category of exposure most service business owners never think about becomes increasingly relevant: employment practices liability.

Claims of wrongful termination, harassment, discrimination, or failure to follow proper employment procedures can be brought against your business regardless of whether the underlying claim has merit. Defending against an employment practices claim — even one you win — can cost tens of thousands of dollars in legal fees.

Employment Practices Liability Insurance — EPLI — covers defense costs and settlements for covered employment-related claims. It's not necessary for a solo operator. As your team grows past three or four employees, it becomes worth discussing with your broker.

The better long-term protection is building the HR practices that reduce the likelihood of valid claims — written job descriptions, documented performance conversations, consistent application of policies across all employees, and clean termination procedures. Chapter 22 covered the documentation habits that protect you. Those records are your first line of defense against employment claims.

Don't Let Coverage Lapse

The most common insurance failure isn't the wrong coverage — it's a lapse in coverage at the wrong moment. A missed premium payment, a policy that renews without you realizing the terms changed, a new service line that falls outside the scope of your existing policy.

Calendar your renewal dates. Review your coverage annually with your broker — not just to confirm it renews but to assess whether your current coverage still matches your actual operations. If you've added employees, expanded your service area, added a new service line, or purchased new equipment, your coverage needs may have changed.

A service business that was adequately covered at $500,000 in revenue may be underinsured at $1.5 million if the underlying policies haven't been reviewed and updated as the business grew. Growth creates new exposures. Coverage needs to grow with it.

The relationship with your insurance broker should be active, not transactional. An annual review conversation — here's what's changed in the business this year, here's what we're planning, does our coverage still fit — is twenty minutes that can prevent a catastrophic gap.

The Legal Agreements That Protect Your Transactions

Insurance protects you from unforeseen events. Legal agreements protect you from foreseeable disputes.

Every significant business relationship should be governed by a written agreement. Not because you assume bad faith — because clear agreements prevent misunderstandings that create disputes that cost money to resolve even when you're in the right.

For recurring service relationships, a simple one-page agreement handles the basics. For larger commercial jobs, a simple scope-of-work agreement signed before work begins establishes what's included, what's excluded, the price, and the payment terms. For employees, an offer letter that confirms the position, compensation, and at-will employment status. For subcontractors, a brief agreement that confirms they're independent contractors, that they carry their own insurance, and that the work is theirs to complete.

None of these need to be complex legal documents. They need to be written, signed, and kept on file. The absence of a written agreement turns every dispute into a he-said-she-said situation that's expensive to resolve and impossible to win definitively.

One clause worth including in any client-facing agreement: a limitation of liability provision that caps your liability for any claim at the value of the services provided. This doesn't prevent claims but it limits the exposure. Many standard service agreements include this language. Have an attorney review your template agreements once — the cost is a few hundred dollars and the protection is ongoing.

Know When to Call an Attorney

Most of what's in this chapter is preventive — the coverage and structure you put in place before something goes wrong. But knowing when to involve an attorney after something goes wrong is equally important.

The threshold for calling your attorney should be low. Any injury on a job site. Any property damage claim above a minor threshold. Any employee termination with unusual circumstances. Any contract dispute that isn't resolved with a direct conversation. Any threat of legal action regardless of how informal.

The instinct to handle these situations yourself — to apologize, to negotiate, to make it right before it becomes a legal matter — sometimes makes sense. Just as often it makes things worse by creating admissions, setting precedent, or agreeing to terms that limit your options later. A fifteen-minute call with your attorney before you respond to a claim or dispute costs very little and can fundamentally change the outcome.

Find an attorney who works with small businesses before you need one. A business attorney in your area who handles contracts, employment matters, and liability claims. Know who they are and have their number saved. The relationship established before a crisis is far more valuable than finding someone during one.

> *Build the protection before you need it. Everything you've built is worth protecting.*

Go Deeper

For more on this and everything else in the book, the Haulers' Edge Newsletter goes deeper every week. Scan the code or subscribe free at HaulingHubb.com.

CH. 29 BUILD IT LIKE YOU'RE GOING TO SELL IT

"The best time to plant a tree was 20 years ago. The second best time is now." — *Chinese Proverb*

Most service business owners never think about selling their business until they're burned out, ready to retire, or facing a life change that forces the decision. By then it's often too late to do the things that would have made the business worth significantly more — because those things take years to build, not months.

The operators who get the best outcomes when they exit aren't necessarily the ones with the biggest revenue. They're the ones who built their business the right way from the start — with documented systems, clean financials, recurring revenue, a team that operates independently, and a customer base that doesn't depend on the owner's personal relationships to stay.

The discipline of building a sellable business doesn't just pay off at exit. It pays off every single day you operate it. A business built to be sold is a business with systems that run without you, financials you can trust, and operations you can step back from without watching it fall apart. Whether you ever sell or not, that's the business worth building.

This chapter is about what makes a service business valuable — and how to build toward that value intentionally starting now.

What a Buyer Is Actually Buying

When someone buys a service business, they're not buying your trucks or your tools or your customer list. They're buying a cash flow stream — the reasonable expectation that the business will continue generating profit after you leave. Everything else is in service of establishing that expectation credibly.

The questions a serious buyer asks are predictable. Does the business generate consistent, documented profit — or does the revenue depend on favorable interpretation of unclear financials? Will the customers stay after the owner leaves — or are the key relationships personal to the current operator? Can the employees execute without the owner's daily involvement — or does everything run through one person's knowledge and judgment? Are there documented systems — or does the business run on institutional memory that walks out the door with the owner? Is the revenue predictable — or is it entirely dependent on new customer acquisition each month?

Every answer that creates uncertainty reduces the value a buyer will pay. Every answer that creates confidence increases it. Every answer that makes a buyer nervous costs you money. Fix those answers now while you have time.

Generally speaking, a business where the owner is essential to operations typically sells for one to two times annual profit, if it sells at all. A business with strong systems, documented processes, recurring revenue, and a team that operates independently can sell for three to five times annual profit or more. On a business generating $200,000 in annual profit, the difference between those multiples is $200,000 to $600,000. The decisions you make today about how to build your operation determine which end of that range you're on.

Clean Financials Are the Foundation

The first thing a serious buyer looks at is your financials. Not your revenue — your profit. And not your stated profit but the documented, verifiable, accountable profit that shows up consistently in your books over multiple years.

Businesses with clean, well-organized financials sell faster and at better multiples than businesses with messy books where the buyer has to guess at true profitability. Businesses where the owner's personal expenses run through the company, where cash transactions aren't recorded, or where revenue and profit vary wildly without explanation don't inspire buyer confidence — and lack of confidence translates directly to lower offers or no offers.

The financial practices that matter for valuation are the same ones that matter for running your business well. Accurate monthly P&L statements. Clean separation of business and personal expenses. Consistent revenue and expense categorization so year-over-year comparisons are meaningful. No unexplained spikes or drops. A clear picture of what's discretionary — the owner perks and one-time items that a buyer would add back to normalize earnings — versus what's core operating cost.

Buyers look at three to five years of financial history when evaluating a business. That means the financial decisions you're making right now are either building or eroding your eventual exit value. Running personal expenses through the business saves a small amount in taxes today and costs a significant amount in valuation later. Clean books consistently over time are one of the highest-return investments you can make in the long-term value of your business.

Work with a CPA who understands small business transactions. Have your books reviewed annually. Know your actual profit margin, not the number you want it to be. The business that knows its numbers with confidence commands more from a buyer than the one where the owner says “well it depends on how you look at it.”

What You Do With Profit Once You Have It

Most business books stop at building profit. This one doesn't, because what you do with profit once you have it determines whether the win compounds or eventually unravels. These five rules apply whether you're planning to sell or planning to run this business for the next twenty years.

First rule: keep a cash reserve before you reinvest or extract. Three months of operating expenses, untouchable, sitting in a separate account. This isn't optional or aspirational — it's foundational. The business that has a cash reserve survives disruptions that destroy unprepared competitors. A bad month, an unexpected repair, a slow season — these are inconveniences when you have a reserve. Without one, they become crises.

Second rule: separate reinvestment from extraction. Some profit belongs back in the business — equipment that improves efficiency, technology that generates more revenue, people who expand what the operation can do. Some profit belongs to you — your personal financial health, your family's security, your ability to build a life outside the business. Both matter and confusing them creates problems. The business that reinvests everything into growth without the owner ever benefiting creates a different kind of burnout than the one that extracts everything and starves the operation of capital.

Third rule: reduce debt intentionally. The loans and leases that were necessary to build the operation become less necessary as cash flow strengthens. Paying down high-interest debt early is one of the highest-return financial decisions available to a service business owner. Every dollar of interest you don't pay is a dollar of profit that stays.

Fourth rule: pay yourself correctly. Not too little, which creates resentment and false financial reporting. Not too much, which starves the business of operating capital. The right salary reflects what you'd pay someone to do your job — the Chapter 6 S-Corp principle applied practically. Your compensation should be explicit, documented, and defensible, not whatever you pull from the business account when you need it.

Fifth rule: think about what you're building toward. Not in the abstract — specifically. Are you building toward a business you'll sell in seven years? That changes how you invest in systems, documentation, and management depth. Are you building toward a business that funds a comfortable retirement without ever selling? That changes how you balance extraction versus reinvestment. Are you building toward something you pass to a family member or a key employee? That changes your hiring and training priorities. The answer shapes every significant financial decision the business makes. Have it.

Recurring Revenue Multiplies Your Value

We spent all of Chapter 27 on recurring revenue. Here's the valuation dimension that chapter pointed toward.

Businesses with predictable recurring revenue are worth more than businesses of identical size that run entirely on new customer acquisition. The reason is risk. A buyer acquiring a business with $50,000 in monthly recurring commercial contracts has more certainty about future cash flow than one acquiring a business doing the same revenue through one-time residential jobs. The recurring business is more predictable, more defensible, and less dependent on continued marketing investment to sustain it.

In practice, a service business with 40% or more of revenue

coming from recurring commercial relationships — property managers, contractors, facility operators on standing agreements — will command a meaningfully higher multiple than an identical business running entirely on transactional residential work.

This doesn't mean residential work has no value. It means building the recurring commercial layer on top of your residential base improves both the stability of your business while you operate it and the attractiveness of your business when you eventually exit. Both outcomes justify the investment in building those relationships.

If you don't have recurring commercial relationships yet, Chapter 27 gives you the playbook. If you have them but they're informal — built on personal trust rather than written agreements — formalize them. A buyer's attorney will ask for documentation of recurring relationships. A handshake arrangement has no value in a transaction. A written service agreement does.

Systems and SOPs: The Owner-Independence Test

The single most important factor in service business valuation — and the one most operators neglect — is whether the business can operate without the current owner.

A buyer is not buying a job. They're buying a business. If the business requires the current owner's personal involvement in day-to-day operations, customer relationships, and decision-making, you don't have a business. You have a job. And nobody buys a job. Buyers discount or walk away from businesses that can't pass this test.

The owner-independence test asks: if the current owner left tomorrow, what would break? The answer tells you everything about what needs to be built before a sale is viable.

If customer relationships are held personally by the owner — clients who call the owner's cell, relationships that haven't been transitioned to systems or other team members — those relationships represent a concentration risk that any sophisticated buyer will price into their offer or use as a reason to lower it.

If the operations run on the owner's knowledge rather than documented SOPs — if the crew knows what to do because the owner trained them personally but that training isn't written down anywhere — the operation is fragile in ways that create buyer anxiety.

If the owner is making daily decisions that could be made by a capable team following clear systems, the business hasn't been systematized to its potential value.

Chapter 20 gave you the SOP framework. Chapter 22 gave you the hiring foundation. Chapter 23 gave you the team development approach. All of that work — the documentation, the training, the systems — directly increases your business's market value by demonstrating that it can continue operating after you step back.

The test to run right now: take a week off. Genuinely step back and let the business run without your daily involvement. What breaks? What decisions come to you that shouldn't? What processes fail without your oversight? Every answer is a gap to close before a buyer finds it.

Customer Concentration: The Risk That Kills Deals

Customer concentration is one of the most common deal-killers in small business transactions. A business where 40% or more of revenue comes from a single client is a fragile business — one client departure changes the financial picture dramatically.

Buyers apply a standard test: if any single customer represents more than 10–15% of total revenue, that concentration is a risk factor that reduces value or triggers specific deal structure protections like earnouts or holdbacks tied to whether key customers stay post-acquisition.

This doesn't mean you shouldn't have large clients. It means you should build enough breadth in your client base that the loss of any single relationship, however painful, doesn't threaten the business's viability. Five recurring commercial clients each representing 10% of revenue is a healthier profile than one recurring client representing 50%, even if the absolute revenue is identical.

If you currently have a concentration problem — a client or two who represent an outsized share of your revenue — the time to address it is now. Not by losing those clients, but by building the base of other revenue that reduces their proportional weight. Every month you spend building additional recurring relationships reduces your concentration risk and improves your eventual exit position.

Equipment, Assets, and Deferred Maintenance

A buyer inherits your assets. Trucks with deferred maintenance, equipment near the end of useful life, and facilities in poor condition are liabilities that a buyer will price into their offer or use as negotiating leverage to reduce the purchase price.

Maintain your equipment properly throughout the life of the business — not just in the year before you sell. Buyers conduct due diligence and they look at maintenance records. A truck with a complete service history commands more in a business sale than an identical truck with no documentation of maintenance.

When you're within a few years of a potential exit, be thoughtful about major capital expenditures. A brand-new truck purchased the year before you sell is an asset that increases the business's book value but may not increase the purchase price by the same amount. Conversely, aging equipment that a buyer will need to replace immediately creates a negotiating point that works against you.

The goal is a business that looks operationally healthy on inspection — equipment maintained, facilities organized, inventory current, no deferred maintenance that signals neglect. A buyer walking through a well-maintained operation has a different emotional response than one walking through a chaotic yard with broken equipment in the corners. Both buy businesses. One pays more.

The Management Layer That Creates Exit Options

A business that requires the owner to be present every day has one exit option: sell to someone willing to replace you as an operator. That's the smallest pool of buyers and they typically pay the lowest multiples.

A business with a management layer — a capable operations manager, crew leads who run jobs independently, administrative staff who handle booking and billing — can be sold to a much wider range of buyers, including passive investors and financial buyers who aren't operators themselves. That wider buyer pool creates competition for your business, which drives up your exit price.

Building toward management independence isn't just about

exit. It's about the lifestyle the business provides while you operate it. Chapter 23 covered building a team that doesn't need babysitting. That team is both your day-to-day quality of life and your exit optionality. The business that can run without you is the business you can sell for the most — and the business you can most enjoy operating in the meantime.

The timeline for building genuine management independence is typically two to three years of deliberate investment in the right people and the right systems. That means the time to start is now, not when you're already thinking about selling.

Tax Structure at Exit

The way you exit matters as much as the price you achieve. Business sales are taxable events and the structure of the transaction — asset sale versus stock sale, installment payments versus lump sum, earnouts, seller financing — has significant tax implications that can affect how much of the purchase price you actually keep.

The difference between a well-structured exit and a poorly structured one on a $500,000 business sale can easily be $50,000 to $100,000 in tax consequences. That difference is entirely determined by decisions made before the sale closes.

Your CPA and your attorney need to be involved in any exit transaction before you've agreed to terms, not after. The time to understand the tax implications of different deal structures is when you still have flexibility to negotiate them. Once a letter of intent is signed and terms are set, your options narrow significantly.

If you're within five years of a potential exit, have a conversation with your CPA specifically about exit tax planning. Strategies like installment sales, opportunity zone investments, and qualified small business stock exclusions exist specifically to reduce the tax burden on business sales. Most operators never access them because they don't know to ask.

Start Building Now

The mistake most operators make is treating business valuation as something to think about later — closer to the exit, when they're ready to sell. By then, the decisions that would have mattered most were made years earlier without the context that would have made them differently.

The financial practices that produce a clean three-year track record need to start three years before you need them. The recurring relationships that demonstrate revenue stability need years of documented history. The SOP documentation that proves operational independence needs time to be built, tested, and refined. The team that can operate without you needs years of investment in hiring, training, and development.

The business that builds toward these standards from day one — not because a sale is imminent but because these practices produce a better business to operate — arrives at any potential exit point in a position of strength rather than a position of catching up.

And if you never sell? You've still built a business with clean financials, documented systems, a strong team, predictable recurring revenue, and the ability to step back without watching it fall apart. That's not a consolation prize. That's the point.

When you're actually ready to sell, the process looks like this: you engage a business broker who specializes in service businesses — expect a commission of 10–12% for transactions under $1M — or approach strategic buyers (competitors, private equity, franchisors) directly. The broker prepares a Confidential Information Memorandum from your financials, runs a controlled process with qualified buyers, and manages due diligence. For a service business under $2M, the sale process typically takes six to twelve months from engagement to close.

The tax strategies worth asking your CPA about at this stage: installment sales spread your gain across multiple years and keep you in lower tax brackets. Opportunity Zone investments can defer and reduce gain if you reinvest in a designated area. None of these are guaranteed to apply to your situation — but each one is worth a twenty-minute conversation before you sign anything.

> *Build it right. Whether you sell it or not, you'll be glad you did.*

Go Deeper

For more on this and everything else in the book, the Haulers' Edge Newsletter goes deeper every week. Scan the code or subscribe free at HaulingHubb.com.

PART 6: FUTURE-PROOFING & AI

A note on this section: The AI landscape is moving faster than any book can track. The principles across these chapters — speed of response, system-building, automation of repetitive tasks, data-driven decisions — are durable and will remain relevant regardless of when you're reading this.

The specific tools, platforms, and capabilities referenced in these chapters reflect the state of the technology at the time of writing — some details will have changed by the time you read this. The strategic principles underlying every chapter in this section will not.

Focus on those.

CH. 30 AI ISN'T COMING — IT'S ALREADY HERE

> *"The machine's danger to society is not from the machine itself but from what man makes of it." — Norbert Wiener*

If you've adjusted a thermostat, followed GPS navigation, or had Google Ads automatically optimize your campaign bids while you were on a job — you've already used artificial intelligence. You just didn't think of it that way.

AI isn't a robot in a lab. It's the invisible layer running underneath the technology you use every day, learning from data and getting better over time. It's been in your business longer than you realize. What's changed isn't that AI arrived — it's that it became accessible enough for a two-truck hauling operation to use it the same way a Fortune 500 company does.

That shift is what makes this moment worth paying attention to.

What AI Actually Is

Most people who feel intimidated by artificial intelligence are intimidated by the name, not the concept.

AI is pattern recognition at machine speed. That's it. Feed a system enough examples, let it find what the examples have in common, and it starts making predictions about new situations it hasn't seen before. The more data, the better the predictions.

Think about a thermostat. It senses the current temperature, compares it to the target, adjusts, and repeats. That loop — sense, decide, act, repeat — is the same loop that powers every

AI system. It's also the same loop every good operator runs. You test a Google Ad headline, see which one converts, and adjust your next campaign. You try a pricing structure, watch the close rate, and refine it. You're already thinking like an AI. The machine just runs the same loop millions of times faster.

Most AI today works by learning from examples rather than following pre-written rules. Feed it enough data, let it find the patterns, and it makes predictions on situations it hasn't seen before. That's what most people mean when they say AI.

Early AI systems were rule-based — engineers wrote specific instructions and the computer followed them. Powerful for structured problems, brittle for anything requiring judgment or nuance. The breakthrough came when researchers stopped trying to program rules and started teaching machines to learn from examples instead. Feed the system thousands of labeled inputs, let it find the patterns, and let it make predictions on new data it hasn't seen.

That's machine learning. And it's what most people mean today when they say AI.

What's Under the Hood

The AI tools that matter most — conversational AI assistants, AI features built into search platforms, and the AI agents in modern CRM platforms like Service Hubb AI and other CRMs — are powered by large language models. Despite the technical name, the concept is approachable.

A large language model is trained on billions of words from books, articles, websites, and conversations. It learns how humans structure sentences, explain ideas, answer questions, and communicate across different contexts. When you type a prompt, it predicts what response should come next based on all that training. That's why it can write a follow-up email, summarize a contract, generate ad copy, or answer a question about your service in a natural-sounding way. It's not thinking. It's predicting — at extraordinary scale and speed.

A simpler frame: your phone's predictive text suggests the next word based on patterns in how you write. A large language model is that, multiplied by a trillion data points. It doesn't have opinions. It has very sophisticated probability tables built from an enormous amount of human communication.

Used with clear direction and appropriate expectations, those probability tables can do things that previously required a specialist — and increasingly they're built directly into the tools service businesses already use.

AI is also not one thing. It's an umbrella term covering several distinct capabilities. Narrow AI does one task extremely well — reading license plates, optimizing ad bids, routing vehicles, predicting customer churn. Conversational AI — like the chat tools you've probably used — handles a wide range of language and reasoning tasks. And increasingly, AI agents combine multiple capabilities with the ability to take sequences of actions toward a goal without step-by-step human instruction. An agent isn't waiting to be told each step. You give it a goal and it figures out how to complete it — checking the calendar, sending the follow-up, logging the update, flagging anything unclear.

That last category is where things are moving fast. And it's already available to you.

What It's Already Doing in Service Businesses

The gap between “AI as a concept” and “AI as something you're already using” is smaller than most people realize.

Marketing. Google Ads has used machine learning to optimize campaign bids in real time for years — deciding in milliseconds which ad to show, to whom, at what bid, based on thousands of signals about the person searching. You're not manually adjusting bids anymore. An AI is doing it constantly based on what's producing results. The operators who understand this adjust their strategy accordingly — they feed the system better inputs, review the outputs critically, and make the strategic calls the AI can't make. The ones who don't are just along for the ride on whatever the algorithm decides.

Lead capture and phone coverage. Missed calls in a service business are lost revenue. AI voice agents now answer every call that your team can't get to — identifying themselves as AI assistants trained on your business, answering common questions, capturing lead information, and ensuring a real person follows up within minutes. We cover the full implementation in Chapter 32. The operators who have deployed this are capturing a meaningful percentage of leads that were previously lost to voicemail. The ones who haven't are still paying for the marketing that generated those calls and then losing them at the last step.

Follow-up and pipeline management. The automated follow-up sequences we built in Chapter 17, the retention campaigns in Chapter 18, the lead pipeline management in Chapter 21 — all of this is AI-assisted automation running in the background while you're doing the actual work. The estimate follow-up that fires on day two, the seasonal reminder that reaches past customers in March, the review request that goes out an hour after a job closes — none of it requires your daily attention once it's configured. It's not magic. It's pattern-based automation built on clear rules. But the output looks like attentiveness and responsiveness to the customer receiving it.

Operations. Routing optimization tools like OptimoRoute apply machine learning to your job list and service windows to produce daily routes that reduce drive time and fuel cost. The analytics in your scheduling platform are identifying patterns in job completion times, no-show rates, and revenue by service type that would take you hours to compile manually. Your accounting software is flagging unusual expenses and cash flow patterns before they become problems.

Content. Before-and-after photos from a job, run through an AI tool, become a social media caption with relevant hashtags in thirty seconds. A voice note describing your week becomes a newsletter draft. A collection of five-star reviews becomes a summary of your competitive advantages for your website. These aren't replacing your judgment and personality — they're eliminating the blank-page problem and handling the mechanical work so you can focus on the direction and approval.

The Equalizer Effect

Here's the shift that matters most for a service business owner operating in 2026.

The capabilities that used to require a marketing department, an operations team, an administrative staff, and a technology team are now accessible through connected software tools at subscription prices a small operator can afford. A solo operator or a two-person crew can run marketing automation, AI-assisted lead capture, automated follow-up sequences, route optimization, and financial analytics with tools that cost a few hundred dollars a month combined. As of this writing, Service Hubb AI — which covers CRM, AI voice and text agents, follow-up automation, review requests, and pipeline management — runs $297 per month with a one-time integration fee of $750 that covers the legal setup for SMS compliance and the full workflow buildout. Add accounting software and a routing tool and you're still well under $500 a month total for the full operational stack.

The large competitor has more trucks and more budget. But if you have better systems — better lead response, better follow-up, better operational efficiency — you win more than your fair share of the market. At Grizzly Junk Pros we are not the cheapest option in our market by a significant margin and we see steady growth every year. The reason is communication. The speed of our response, the clarity of our follow-up, the consistency of how we handle every touchpoint — customers feel the difference even if they can't name it. Systems and speed beat size when the customer is making a real-time decision. The technology that was an advantage exclusive to large operators has become infrastructure available to everyone.

That's the window that's open right now. The operators who build these systems in the next two years will have operational advantages that compound. The ones who wait will spend years catching up.

AI and Jobs: What's Actually Happening

The conversation about AI and employment is real and worth engaging honestly rather than either dismissing or catastrophizing.

AI replaces tasks, not jobs — initially. But most jobs are a collection of tasks, and when the majority of those tasks can be done by AI, the job changes. The administrative work that required a dedicated person — scheduling, invoicing, follow-up, basic customer communication — increasingly gets absorbed by connected systems. Before we had this running at Grizzly Junk Pros, we were spending ten to fifteen minutes of admin per job. At thirty jobs a week that was seven to eight hours — a full workday — lost to paperwork every week. Automation gave that time back. What used to take a part-time admin can often run on automation. What used to take a dispatcher can run on routing software with occasional human oversight.

This isn't happening slowly. It's happening in the time frame of a business cycle. The businesses that understand this and build their operations around it will run leaner and more profitably. The ones that don't will carry cost structures that their AI-enabled competitors don't.

For the service business owner reading this book, the practical implication is clear. Build your systems now. The time you invest in setting up automation, configuring AI tools, and documenting your processes into repeatable workflows is time that pays back indefinitely — not just once.

The human elements that AI genuinely can't replace are the ones worth doubling down on. The relationship with the contractor who's been calling you for three years. The judgment call on a difficult job that doesn't fit any standard scenario. The trust a customer extends when they give you access to their home. The accountability that comes with your name on the business. AI enhances these things by handling everything around them. It doesn't replace them.

Your customers don't want to talk to an algorithm when something goes wrong. They want a real person who cares and can make a decision. That's your permanent competitive advantage in a world where everything else is getting automated. Protect it and invest in it alongside your technology stack.

Use It Wisely

AI makes mistakes. It can misread tone, miss nuance, hallucinate facts it presents confidently, and inherit biases from the data it was trained on. None of these are reasons to avoid it. They're reasons to use it with oversight rather than on autopilot.

The frame that works is the same one you'd apply to a capable new employee. You don't hand them the company credit card and send them off unsupervised. You give them clear direction, review their work, correct their mistakes, and expand their responsibility as they demonstrate they can handle it. AI works the same way. Clear prompts produce better outputs. Reviewed outputs catch errors before they reach customers. Refined instructions over time produce increasingly reliable results.

Data security and privacy are real considerations. Be thoughtful about what information you feed into AI tools and ensure you're using reputable platforms with clear data policies. For most operational uses — marketing, scheduling, follow-up, routing — the tools available from established platforms operate with appropriate safeguards. For anything involving sensitive customer information, verify the platform's data handling practices before connecting it.

The businesses that will get the most from AI in the next decade aren't the ones who adopt every new tool the week it launches. They're the ones who identify the highest-leverage applications for their specific operation, implement them with appropriate oversight, and build the operational habits that make AI an asset rather than a liability.

Here's what that oversight looks like in practice. When your AI voice agent handles calls, listen to three recordings within 24 hours of going live. Check: did it capture the right information? Did it answer accurately? Did it escalate when the caller was frustrated? Fix one thing in the training prompt. Repeat weekly for the first month. After 30 days of weekly refinement, move to spot-checking one call per week. That's it — that's the oversight system. It takes twenty minutes a week in month one and five minutes a week after that.

You Don't Need to Be Technical

The single biggest barrier most service business owners feel when they encounter the AI content in this book isn't time or money. It's the assumption that this requires a technical background they don't have. It doesn't.

You don't need to know how to code. You don't need to understand machine learning or software architecture or APIs. You don't need to be anything other than someone who can describe what they need in plain language — because that's exactly how these tools work. You talk to them the way you'd talk to a capable employee. You describe the problem, ask the question, or explain what you're trying to accomplish. The AI figures out the rest.

Here's the part that makes this genuinely different from every other technology shift in business history: you can use AI to learn how to use AI. If you don't understand something in this book, open a chat with any of the major AI language models available right now and ask it to explain the concept to you in plain language. Ask it to walk you through a specific setup step by step. Ask it to give you an example using your specific business. It will. And it will keep explaining, rephrasing, and simplifying until it lands — without judgment, without impatience, and without charging you a consulting fee.

The tools themselves are changing rapidly and the landscape will look different by the time you read this than it did when it was written. What won't change is the fundamental capability — AI as a patient, knowledgeable resource you can access at any hour to learn whatever you need to learn next.

And the learning capability goes further than most people realize. Today you can take a four-hour YouTube video, a full-length business book, a long-form podcast transcript, or any piece of long-form content and feed it directly into an AI platform. Tools built specifically for this — designed to ingest large amounts of content and let you interact with it — have made something possible that simply didn't exist until very recently. Instead of watching the whole video or reading the whole book, you can ask the AI to extract the specific insights that apply to your business, summarize the frameworks in plain language, and show you exactly how to implement the relevant ideas in your specific context. A book that would have taken you two weeks to read and another month to figure out how to apply can now be processed, filtered, and implemented in a single focused session.

You can take this even further. You can train an AI on your own business — your SOPs, your pricing model, your customer communication style, your service area, your team structure — and then ask it to apply the teachings from any book, course, or framework directly to your operation. Not generic advice. Advice filtered through the specific reality of how your business runs. That used to cost you five grand from a consultant. Now you do it yourself in an afternoon.

Which brings us to the most valuable skill you can develop — not just for AI but for everything that follows it. The ability to learn. Specifically the ability to move from not knowing something to implementing it faster than you ever thought possible. That skill — learning how to learn better and faster — is what separates the operators who stay ahead from the ones who are always catching up.

Not long ago, going from zero knowledge on a topic to functional implementation took months. You'd find a course, read a book, hire a consultant, attend a seminar. The timeline from curiosity to capability was measured in weeks at best. Today that same journey can happen in days instead of months. You identify what you need to know, you ask an AI to teach you, you ask it to walk you through the steps, you ask it to troubleshoot when something doesn't work. The feedback loop that used to take weeks now takes minutes.

That compression — from not knowing to doing at a speed that would have seemed inconceivable just a few years ago — is the real opportunity sitting in front of every service business owner reading this book. The operators who internalize that and build the habit of rapid learning and rapid implementation will compound their advantage in ways that are genuinely difficult for slower-moving competitors to overcome. Not because they're smarter. Because they got comfortable moving fast from unknown to done.

You're technical enough. You just have to start. Here's the starting point — one exercise that takes five minutes and proves the concept. Open ChatGPT, Claude, or Gemini on your phone. Type: "I run a [your service] business in [your city]. What are the three most common reasons service businesses in my trade lose money on jobs?" Read what comes back. Then ask: "Now give me a five-question checklist I can use before quoting every job to make sure I'm not underpricing it." You just used AI to build a business tool in under five minutes. That's the starting point for everything else in Ch. 31.

Where This Is Going

The trajectory of AI improvement has been consistent — each year brings capabilities that were unavailable the year before. That pace shows no signs of slowing.

AI today is still early enough that operators who build proficiency now will have a meaningful head start — the same way businesses that figured out smartphones early had advantages that compounded for years. The platform is established, the core capabilities are proven, and we're just beginning to discover how to use them well in specific operational contexts.

The same dynamic applies here. The window to build genuine proficiency before your market reaches saturation is shorter than most people expect. The next chapter gets into the specific AI applications for service businesses. Chapter 33 covers the tools that are already operating with a level of autonomy that most people don't realize is available. And Chapter 35 makes the case directly for why your next competitor may come from a background you wouldn't expect — precisely because they understand how to deploy these tools.

The operators who adapt early will pull ahead. The ones who wait will spend years catching up on someone else's terms. That dynamic applies whether you're reading this in the year it was published or five years later — the gap between operators who build and operators who don't only widens over time.

AI won't take your customers. But a competitor using AI well might.

Understand it. Use it. Direct it. That's the edge.

Go Deeper

For more on this and everything else in the book, the Haulers' Edge Newsletter goes deeper every week. Scan the code or subscribe free at HaulingHubb.com.

CH. 31 HOW TO USE AI BEFORE YOUR COMPETITOR DOES

> *"The greatest danger in times of turbulence is not the turbulence — it is to act with yesterday's logic." — Peter Drucker*

When I started in 2014, answering every call was a competitive advantage. Most companies in my market were slow, inconsistent, and hard to reach. We built a culture around responsiveness — if we missed a call, our voicemail promised a callback within thirty minutes and we lived by it. That standard alone won jobs.

That edge is gone. Responsiveness is now the floor, not the ceiling. Online booking, active reviews, professional presentation — all of it is expected before a customer will even consider calling you. The bar keeps rising and what differentiated you two years ago is table stakes today.

The next layer of advantage is AI — and most service businesses haven't built it yet. That gap is the opportunity. The operators who deploy these tools now, while the window is still open, will have systems in place that their competitors are scrambling to catch up to in two years.

This chapter is the operational playbook. Not theory — specific applications, in the order that matters most, with the guardrails that keep them working.

Why the Timing Matters

Studies across service industries put the average missed-call rate for small businesses at roughly 62% — your number may be better or worse depending on how consistently you or your team answer *(research widely cited across service industry studies)*. And 85% of those callers never leave a voicemail or try again. They call the next company on the list.

If your average job is worth $350 and you're missing even one lead a day, that's over $100,000 a year in potential revenue — even at a 30% close rate on those recovered calls, that's $30,000 a year walking out the door through a problem that's entirely solvable.

Speed of response compounds this — as covered in Chapter 16, the lead who doesn't hear back within minutes is already calling someone else. Most service businesses are responding hours later — or not at all.

AI doesn't fix your service. It doesn't replace your crew or your judgment. What it does is close the operational gaps where revenue is currently leaking — missed calls, slow quotes, forgotten follow-ups, manual admin — so more of the work you're already generating actually converts into booked jobs and paid invoices.

The tools to do this are available now, affordable at small business scale, and integrating directly into platforms most service operators are already using or should be. Here's where to start.

Answering Every Call

Every call that goes to voicemail is a lead you paid to generate that you're handing to your competitor. After hours it's worse — most of those calls simply die.

The fix is an AI voice agent that answers every call your team can't get to, identifies itself as an AI assistant trained on your business, and captures the lead before the caller moves on.

In practice it looks like this. A prospect calls your number. If you don't answer within two rings, the AI picks up — "Hey, thanks for calling Grizzly Junk Pros. I'm a virtual assistant here to help capture your info so the team can follow up fast. What are you looking to get rid of? What's the zip code?" If the caller hangs up before leaving information, the system immediately sends a text — "Got your call — feel free to text us photos for a fast quote." Either way, the details flow into your CRM, you get a notification, and the customer gets a response instead of silence.

The first week we had this running, the system booked seven jobs I would have missed. Roughly $2,500 in recovered revenue in seven days without me touching the phone once.

As covered in Chapter 16, the principle that makes this work is transparency — the agent identifies itself as AI immediately and clearly, and a real person follows up fast. Honesty here isn't just ethics — it's what makes the system perform.

After hundreds of calls through our AI agent, I can count on one hand the number of callers who even asked whether they were talking to a human. The system sounds natural, gets more natural over time as you refine it, and keeps improving as you review recordings and adjust the training.

The guardrails that make this work: train the agent on your actual FAQs, service area, pricing structure, and what you won't take. Set clear escalation rules — the moment a caller is upset or has a complex situation, route to a human immediately. And review call recordings regularly in the early weeks to catch anything that needs refinement.

Service Hubb AI handles this natively — the voice agent integrates directly with your lead pipeline so every captured call creates a record and triggers your follow-up sequence without manual input.

Faster Quotes

A customer texts you photos of the job while you're hauling.

The photos sit in your messages for hours. By the time you get back to them, they've booked someone else.

Speed of estimate delivery is one of the highest-leverage conversion factors in a service business. Most operators take 24 to 48 hours to quote. The businesses that quote within thirty minutes win a disproportionate share of the jobs.

AI-assisted quoting closes that gap. The customer sends photos through your website form, text, or booking link. The AI analyzes the images, checks your pricing rules — disposal fees, surcharges, access conditions — and generates a draft quote. You get an alert, review it in two minutes, adjust if needed, and approve. The customer has an estimate in their inbox while their intent is still hot.

For junk removal we use Haulers' Edge AI for this function. For other service trades, similar photo-to-estimate tools are available or being built into field service platforms. The specific tool matters less than the principle: any system that dramatically compresses the time between customer inquiry and quote delivery improves your conversion rate.

Keep a human in the loop for every quote, especially early in the implementation. Provide price ranges rather than exact numbers when the job has ambiguity — "likely between $200 and $280 depending on what we see on-site." Set your AI pricing rules to account for your actual local costs, not national averages. And when uncertain, estimate slightly high — it's easier to come down on-site than to explain a bill that exceeded the estimate.

Follow-Up That Runs Itself

Most jobs aren't lost on the first call or even on the estimate. They're lost in the silence that follows.

You send a quote, get buried in the day's work, and never circle back. The customer moves on — not because your price was wrong but because nobody followed up. The data behind this — covered in full in Chapter 17 — is consistent. That gap between what the data says and what operators actually do is where jobs slip away.

Automated follow-up sequences run the three-touch cadence from Chapter 17 without you managing it — written in your voice, timed automatically, stopping the moment the customer responds or books.

A single additional follow-up improves close rates meaningfully. Three well-timed touches stack that improvement further. You're not pestering — you're showing up. The customers who eventually book after a third follow-up are customers you would have permanently lost without it.

Configure this in Service Hubb AI as a pipeline automation — when an estimate is marked sent, the sequence starts. When the customer books or responds, it stops. Set it once and it runs on every estimate indefinitely.

Routing That Pays for Itself

The dispatch system most service businesses run on is a combination of a whiteboard, a mental map of the service area, and the order jobs came in. The result is trucks crossing paths all day, unnecessary miles, fuel waste, and crew arriving at the last job exhausted.

AI-assisted routing optimization takes tomorrow's job list — addresses, time windows, rough durations, dump run requirements — and generates an optimized sequence in seconds. Jobs cluster geographically. The dump run happens at the right point in the day rather than as an afterthought. The crew runs a logical loop instead of zigzagging.

The first time we ran an optimized schedule we cut nearly fifty miles in a single day. Studies consistently show route optimization reduces drive time and fuel consumption by 20–30% *(American Transportation Research Institute, 2023)*. On a fleet running five days a week, that's thousands of dollars annually in recovered fuel and labor costs — plus the occasional extra job that fits in the day because the routing is tight enough to create a window.

Chapter 19 covers the full routing strategy. The tools that execute it — OptimoRoute, Route4Me, or the routing features built into your field service platform — are the operational layer that makes the strategy automatic rather than dependent on someone building an optimal sequence manually each morning.

Build in buffers. The algorithm doesn't know that a particular customer always needs an extra fifteen minutes or that a specific neighborhood has a school zone that creates a noon backup. Your drivers do. Use optimization to build the plan and use driver knowledge to refine it.

Admin That Closes Itself

Before automation, the workday ended when the truck came back but the work didn't. Invoicing, payment follow-up, review requests, expense logging, CRM updates — another hour of admin for every day on the road.

Connected workflow automation closes this. When a job is marked complete in your system, a sequence fires automatically. The invoice generates and sends. If a card is on file it processes immediately. A thank-you text goes to the customer with a direct link to leave a Google review. Job photos, signatures, and notes get stored to the customer record. A follow-up reminder for seasonal work queues up for the right time of year.

One click closes the job and the entire back end handles itself.

Before we had this running, we were spending ten to fifteen minutes of admin per job. At thirty jobs a week, that was seven to eight hours — a full workday — lost to paperwork every week. Automation gave that time back. More importantly, nothing slipped. Every job billed. Every review request went out. Automated review requests doubled our monthly review volume, which compounded into better local search ranking, which drove more inbound leads.

Keep oversight on exceptions — large refunds, discounts, unusual situations. Space the customer communications logically — invoice first, then receipt, then review request, not all at once. Write the automated messages in your actual voice, not corporate template language. The customer should feel like they heard from you, not from a system.

Getting Your Team On Board

When you introduce AI to your operation, expect skepticism before you get buy-in. Your crew hears "automation" and thinks "replacement." That's a natural reaction and it's your job to reframe it before resistance hardens.

The framing that works is capacity, not efficiency. AI takes the repetitive administrative work off everyone's plate so the humans in the business can focus on the work that actually requires them — the customer relationship, the physical job, the judgment calls. It doesn't replace what your crew does. It handles everything around what they do so they can do it better.

Involve them early. Before rolling out a new tool, explain what it does and why. Ask for their input on what's working and what's creating friction. When people feel heard they're more likely to engage with the change rather than resist it. When a driver tells you the routing software suggested a sequence that doesn't account for a specific traffic pattern they know about, update the system. That feedback loop makes the tool better and signals that their experience matters.

When the system wins — when a job gets booked from an after-hours call that would have been missed, when the routes run tight and the day ends early — name it. "The routing cut thirty minutes off today's schedule." "The AI caught a follow-up that would have gone cold." Small visible wins build the team's confidence in the system faster than any internal roll-out presentation.

What Good Looks Like

One objection worth addressing before the specifics: if you've watched business books recommend software that was obsolete two years later, you're right to be skeptical of tool-specific advice. This chapter focuses on capabilities and outcomes rather than brand names for exactly that reason. The specific tools will evolve — some will consolidate, some will be replaced by better options, some are already changing as this is written. What won't change is the underlying capability: faster lead response, automated follow-up, intelligent routing, admin that closes itself. Build toward those capabilities. The tools that deliver them today are the starting point, not the destination.

AI is only as valuable as what it actually produces. The benchmarks that tell you whether your implementation is working:

A 95% or better answer rate on inbound calls and messages. If leads are still hitting voicemail consistently, the coverage isn't working.

Quotes delivered to 80% of inquiries within thirty minutes of the customer sending details. If you're still taking a day to quote, you're losing the speed advantage.

Three to five follow-up touches landing on every unbooked estimate. If prospects are going dark and you're not following up, the sequence isn't running.

Route mileage ten to twenty percent below your pre-optimization baseline. If the trucks are still zigzagging, the routing isn't being used or isn't configured correctly.

Two to three new reviews per week arriving without manual effort. If reviews aren't compounding, the review request automation isn't firing.

Track these numbers. Before you implement anything, log your current baseline on each metric. After sixty days, compare. If the numbers aren't moving, something in the configuration is wrong. If they are, you have evidence to justify expanding the investment.

The Experience From the Customer's Side

Here's what the fully implemented version looks like from a customer's perspective.

It's 9:47pm and they message your Facebook page asking for a quote. Normally that would sit until morning. Instead they get an immediate response — a few friendly questions, a prompt to send photos. Within minutes they have an estimate and a booking link. Two other companies they messaged haven't replied yet. You've already won the first impression.

After booking they get a confirmation and prep instructions. The morning of the job they get a text window for arrival. Twenty minutes out another text. After the job, a thank-you and a one-tap review link while they're still feeling good about the experience. They leave a five-star review mentioning how smooth everything felt.

They didn't know AI was involved. They just thought your company has its act together.

That's the goal. AI should be invisible to the customer — what they experience is responsiveness, clarity, and follow-through. The technology is the mechanism. The experience is what they remember and tell people about.

Test it yourself. Send an inquiry to your own business at 10pm. How fast does a response come? Does it sound like you? Is the follow-up timed right? Find the friction points before your customers do and eliminate them.

> *Speed plus consistency equals profit. Build the system that delivers both.*

Go Deeper

For more on this and everything else in the book, the Haulers' Edge Newsletter goes deeper every week. Scan the code or subscribe free at HaulingHubb.com.

CH. 32 THE EMPLOYEE THAT NEVER CALLS OUT SICK

> *"The first rule of any technology used in a business is that automation applied to an efficient operation will magnify the efficiency. The second is that automation applied to an inefficient operation will magnify the inefficiency." — Bill Gates*

Chapter 31 made the case for AI voice agents and showed you what they produce. This chapter is the build guide — how to configure, train, and run one that captures revenue around the clock without damaging the customer experience.

What an AI Voice Agent Actually Does

An AI voice agent is software that answers phone calls and engages the caller in a natural, conversational way — asking questions, providing information, and capturing details — without a human on the other end.

The distinction that matters for a service business is between an AI voice agent and an interactive voice response system — the "press 1 for sales, press 2 for support" menus that everyone hates. Those systems are rigid, scripted, and impersonal. A well-configured AI voice agent sounds like a conversation. The caller asks a question, the agent responds naturally. The caller says they have a garage full of old furniture and appliances, the agent asks follow-up questions — what zip code, roughly how much, is there anything oversized — and builds a picture of the job.

The goal isn't to replace human conversation. It's to ensure that when a human can't be there, something intelligent, responsive, and trained on your business picks up instead of voicemail.

There are two scenarios where the AI voice agent earns its keep every day.

Overflow during business hours. You're on a job, your admin is on the phone, and another call comes in. Without an AI agent, that call goes to voicemail. With one, it gets answered immediately by an agent that knows your business and starts capturing the lead while you finish your current conversation.

After-hours coverage. Every call that comes in outside your business hours — evenings, weekends, early mornings — gets answered with the same quality of response as a call during peak hours. The lead gets captured. A notification fires to the right person. Follow-up happens within minutes of the next business window opening rather than never.

The Non-Negotiable: Be Transparent

We covered this principle in Chapter 16 — it's the single most important configuration decision you'll make. The agent must identify itself as AI clearly, at the very beginning of the call. Something like: "Hi, thanks for calling Grizzly Junk Pros. I'm an AI assistant trained on everything about the business — I can answer your questions, get your details, and make sure the right person follows up with you fast."

Be transparent. It works better anyway.

Building Your Agent in Service Hubb AI

Service Hubb AI includes AI voice and text agent functionality built specifically for service businesses. The setup process involves four components: training, scripting, escalation rules, and pipeline connection.

Training — what the agent knows. Before your agent takes a single call, it needs to know your business thoroughly. This isn't a quick setup. It requires deliberate input and it's the work that determines whether the agent sounds competent or confused.

Start with your service area. Every zip code you serve, every zip code you don't. If a caller is outside your area, the agent needs to know to say so clearly rather than capturing a lead that can't convert.

Next, your services. Every service you offer, how you describe it, what it costs in general terms, and what it doesn't include. If you rent 20-yard dumpsters only, the agent should know that and be able to explain it. If you don't take hazmat, tires, or certain electronics, the agent needs to know that and communicate it without confusion.

Your pricing structure. Not necessarily exact prices, but ranges by service type that give the caller enough information to understand whether you're in their ballpark. "Junk removal typically runs between $200 and $600 depending on the volume and what's included — the team can give you an exact number once we know more about the job" is an honest, useful answer. "I can't give you any pricing information" is not.

Frequently asked questions. Pull these from your own experience — the questions you answer ten times a week. Are you licensed and insured? Do you recycle or donate? How quickly can you schedule? Do you offer same-day service? What happens if the job is bigger than expected? Build the agent's answers from your actual responses, not generic customer service language.

What you won't take. Be specific. The caller who wants to get rid of paint, chemicals, or medical waste needs a clear answer early in the conversation rather than after they've booked a job.

Scripting — how the agent talks. The agent's voice should sound like your business, not like a generic customer service bot. If your brand is casual and direct — which it should be based on everything this book has built toward — the agent's language should match. Short sentences. Plain words. A tone that feels like a knowledgeable team member, not a corporate FAQ page.

Write your opening, your primary qualifying questions, your common objection responses, and your close — the part where the agent explains what happens next. "I've got everything I need — I'm sending this to the team right now and you should hear back within the hour. Is there anything else I can help with before I let you go?"

Read every line out loud before you finalize it. If it sounds like something a real person would say, keep it. If it sounds like a script, rewrite it.

Escalation rules — when to hand off to a human. Not every call should stay with the AI agent. Configure clear triggers that route to a live person or generate an immediate alert.

A caller who is upset or frustrated should reach a human as soon as the agent detects that signal — through word choice, tone, or an explicit statement. An AI agent responding to an angry customer with calm scripted responses is not a good customer experience.

A caller asking a complex question the agent isn't trained to answer should be offered a callback rather than a guess. "That's a great question and I want to make sure you get an accurate answer — can I have someone call you back within the hour?"

A high-value commercial lead — a property manager calling about a standing account, a contractor calling about ongoing dumpster service — should get flagged for immediate human follow-up regardless of whether the agent captures the basics.

And any caller who explicitly asks to speak with a human

should be accommodated. The agent can explain that the team is currently unavailable and commit to a specific callback window — but it should never try to retain a caller who wants a person.

Pipeline connection — where the lead goes. This is what separates a functional AI agent from a sophisticated voicemail. When the call ends, the information captured should flow directly into your Service Hubb AI pipeline — creating a lead record, tagging the source, and triggering an automated notification to whoever needs to follow up.

If the call comes in at 11pm, the on-call person gets a text notification with the lead details. The follow-up sequence starts. By the time the caller wakes up the next morning, they may already have a message from your business. That speed — follow-up happening within minutes even on an after-hours lead — is what makes the difference between capturing the job and losing it to whoever called back first.

The Text Agent: Same Principle, Different Channel

Everything in this chapter applies equally to text-based AI agents. Many customers — especially younger demographics and commercial clients who prefer written communication — will text your business number rather than call.

A text agent that responds immediately to incoming messages, asks qualifying questions, captures lead information, and connects to your pipeline produces the same revenue recovery effect as the voice agent on a different channel.

The 9:47pm Facebook message scenario from Chapter 31 — the customer who messages after hours and gets an immediate intelligent response while two competitors go dark — is a text agent application. Configure Service Hubb AI to handle inbound texts through the same training and pipeline connection as your voice agent. The customer experience is consistent across channels. The lead capture is automatic. The follow-up is triggered regardless of when the inquiry arrives.

Reviewing and Refining

The agent you launch on day one is not the agent you'll have at day thirty. Continuous refinement is what turns a functional AI agent into a genuinely effective one.

Review call recordings regularly in the first month. Listen for moments where the agent's response was confusing, off-brand, or unhelpful. Fix each one specifically — update the training, adjust the scripting, add a FAQ the agent didn't have the right answer to. Every refinement compounds over the hundreds of calls that follow.

Track the metrics that tell you whether the agent is working. Answer rate on calls that would have gone to voicemail — you want this above 95%. Lead capture rate from agent calls — what percentage of callers are providing enough information to become a qualified lead record. Follow-up conversion rate — of the leads captured by the agent, what percentage book a job. If leads are being captured but not converting, the follow-up process needs attention. If the answer rate is high but lead capture is low, the agent's qualifying questions need refinement.

Ask your team what they're hearing from customers. If you're getting callbacks that start with "I talked to your automated system and it wasn't sure about..." — that's a training gap to close. If customers are mentioning that the after-hours response impressed them — that's confirmation the system is delivering the experience it should.

What It Costs Versus What It Recovers

The math on AI voice agent implementation is straightforward when you run it honestly.

Service Hubb AI's pricing includes the AI voice and text agent functionality as part of the platform. The total cost of the platform — CRM, pipeline management, automation, AI agents — is a fraction of what a part-time admin would cost to cover the same functions.

Against that cost, set the revenue recovery. If your average job is worth $350 and your AI agent captures one job per week that would otherwise have gone to voicemail — conservative for any operation getting more than twenty inbound calls per week — that's $1,400 per month in recovered revenue. Two jobs per week is $2,800. The math doesn't require a large capture rate to produce a compelling return.

The non-revenue benefits compound on top. Every job captured after hours is a job your competitors don't get. Every caller who experiences immediate response rather than voicemail forms a different first impression of your business. Every lead that flows into your pipeline automatically rather than being manually logged is time returned to your team.

The employee that never calls out sick, never needs a raise, never has a bad day, and never lets a call go unanswered works every hour of every day for a cost that most service businesses spend on two or three tanks of diesel per month.

Build it.

> *The call you miss tonight is the job your competitor books tomorrow. Close the gap.*

Go Deeper

For more on this and everything else in the book, the Haulers' Edge Newsletter goes deeper every week. Scan the code or sub-

scribe free at HaulingHubb.com.

CH. 33 YOUR FIRST DIGITAL EMPLOYEE

"Before you automate, simplify." — *Elon Musk*

This chapter is for the operator who's implemented the tools in the last two chapters and is ready to go further.

There's a version of your business where a significant portion of the research, prospecting, outreach preparation, and operational intelligence work gets done without you touching it — not because you hired someone to do it, but because you built an AI agent that does it while you sleep.

That's not a distant vision. It's available today — and the barrier to entry is lower than you might think.

This chapter is about agentic AI — AI that doesn't just answer questions but takes sequences of actions toward a goal. It's about Claude Code specifically, which emerged as a genuinely transformative tool after this book was originally written and deserves its own chapter. And it's about the practical applications that matter most for a service business: building lead lists, automating research, and creating operational leverage that compounds over time.

The Difference Between AI Tools and AI Agents

Most of the AI covered in previous chapters — the voice agent, the follow-up sequences, the quoting tool — operates reactively. A customer calls, the voice agent responds. An estimate gets sent, the follow-up sequence triggers. You ask ChatGPT a question, it answers.

Agentic AI operates proactively. You give it a goal — not a question, a goal — and it figures out the steps required to accomplish it, executes those steps in sequence, makes decisions along the way, and delivers a result. You're not in the loop for every action. You set the objective and review the output.

The distinction matters because it changes what's possible. Reactive AI makes your existing workflows faster. Agentic AI creates entirely new workflows that didn't exist before because they required human time and judgment that was previously too expensive to deploy at scale.

A simple example. You want a list of every general contractor in your service area with their company name, phone number, website, and estimated project volume. Compiling that manually would take days — searching Google, visiting websites, cross-referencing databases, building a spreadsheet. An AI agent given that goal can execute the research, structure the data, and deliver a formatted CSV while you're running jobs. The research that would have taken you or an admin a full day takes the agent a few hours of unattended work.

That's what your first digital employee does.

Claude Code: What It Is and Why It Matters

Claude Code is Anthropic's agentic coding tool — a command-line interface that gives Claude the ability to not just generate code but execute it, access files, run terminal commands, browse the web, and complete multi-step tasks autonomously. It's built for developers but its capabilities extend far beyond coding into research automation, data collection, and operational workflow building.

I've been running Claude Code research agents on a Mac Mini M4 in my office. The setup is straightforward — the Mac Mini runs headlessly, accessible via remote desktop, with Claude Code configured to execute research tasks and save outputs to a designated folder. I give it a target — contractors, roofers, property managers, real estate investors in my service area — and it builds lead lists as CSVs that feed directly into my outreach workflow.

The specific value for a service business comes from three capabilities that weren't practically available before Claude Code.

Automated lead research. Give the agent a target customer profile — "general contractors in [county] who pull more than 10 permits per year" — and it can query public permit databases, cross-reference business directories, visit company websites to confirm contact information, and deliver a structured lead list. This is research that previously required either significant staff time or expensive data vendors. Claude Code does it for the cost of an API subscription.

Competitive intelligence. Give the agent a list of competitors and it can systematically gather their pricing (where public), their review profiles, their service offerings, their geographic coverage, and their marketing messaging. The kind of thorough competitive analysis from Chapter 7 that might take an afternoon of manual work can run automatically on a regular schedule.

Content and operational research. Give the agent a topic — seasonal demand patterns in your market, pricing benchmarks for a specific service type, emerging regulations affecting your business — and it researches, synthesizes, and summarizes. This is the research function that most small operators either skip or do superficially because it's time-consuming. An agent does it thoroughly and consistently.

Setting Up Your Research Agent

A note before the setup instructions: this chapter said earlier that you don't need to be technical — and that's true for the daily operation once the agent is running. But the initial setup does require opening a computer terminal, running a few installation commands, and doing some basic configuration. If you've never used a terminal, that will take an afternoon of learning, not five minutes. Two options for getting past that: work through Anthropic's setup documentation step by step and use Claude itself to troubleshoot any error messages you hit — paste the error into a Claude chat and ask what it means. Or hire someone to handle the installation and configuration while you handle the prompt design and output review. Either way, after setup the daily operation is straightforward. The technical barrier is real but it's a one-time door, not a permanent wall. Windows users: Claude Code runs on Windows as well — the specific commands differ slightly from the Mac instructions below, and Anthropic's documentation covers both.

You don't need to be a developer to build and run a basic Claude Code research agent. You need a Mac or PC that can run continuously, a Claude Pro or Max subscription, and a willingness to spend a few hours on initial configuration. Here's the practical setup.

The hardware. A Mac Mini is what I use — quiet, energy-efficient, powerful enough to run AI agents continuously, and small enough to sit in a corner of an office or warehouse. It doesn't need to be on your desk or even in your primary workspace. It just needs to be powered on and connected to your network. A Windows machine works equally well if that's what you have.

The software. Claude Code runs from the terminal — the command-line interface on your computer. On a Mac, access it through Terminal or iTerm2. The installation is a few commands that Anthropic's documentation walks you through. Once installed, you run claude from the terminal to start a session. One configuration note specific to Mac: run source ~/.zshrc before launching Claude in each new terminal session to ensure your environment is configured correctly.
The workflow. You open a Claude Code session, give it a de-

tailed prompt describing the research task and the output format you want, and let it run. Research outputs save to a designated folder — I use /Research/ on the Mac Mini — as CSVs or structured text files that I can open in Excel or import directly into Service Hubb AI as lead lists.

Remote access. If your research machine isn't at your primary workspace, you access it through Screen Sharing on Mac or Remote Desktop on Windows. This lets you start a research session, check progress, and retrieve outputs from anywhere — from your phone, your laptop, or another computer.

The Lead Lists That Feed Your Outreach

The most immediately valuable application for most service businesses is automated lead list building — creating targeted prospect lists for outreach without manual research.

For Grizzly Junk Pros, the target audiences that produce the best recurring revenue are contractors, roofers, property managers, real estate investors, and estate attorneys. Each of those categories has public data available — licensing databases, permit records, business directories, professional association member lists — that can be systematically gathered and structured by an AI agent.

A well-written prompt for a lead research agent includes: the target customer type, the geographic parameters, the specific data fields you want in the output (company name, contact name, phone, email where available, website, any qualifying criteria), the sources you want the agent to check, and the output format.

The more specific your prompt, the more useful the output. "Contractors in [county]" produces a broad list. "General contractors in [county] who have pulled residential renovation permits in the past 12 months and have an active website" produces a qualified list of people who are actively doing the work that generates demand for your service.

The CSV the agent delivers goes directly into Service Hubb AI as a prospect list. Your outreach sequence — the email and text cadence from Chapter 17, adapted here for cold prospects — runs on that list automatically. The agent builds the list. The CRM runs the outreach. You follow up on the responses. The entire top-of-funnel process runs without your daily involvement once it's configured.

Beyond Lead Lists: What Else Your Digital Employee Can Do

Lead research is the entry point. The capability extends further as you get comfortable with what's possible.

Market monitoring. An agent configured to check competitor Google Business Profiles, review counts, and pricing pages on a weekly schedule gives you the competitive intelligence from Chapter 7 running automatically rather than manually. You get a weekly summary of changes rather than having to remember to check.

Content research. An agent that researches what questions your target customers are asking on forums, in Facebook groups, and in review responses gives you an endless supply of content ideas for your blog, your newsletter, and your social media — all drawn from what real customers in your market are actually asking.

Regulatory and operational research. Dump fee changes, new disposal regulations, permit requirement updates, licensing changes in your state — all of these affect your operations and pricing. An agent monitoring relevant government websites and industry publications surfaces these changes before they become surprises.

Customer data enrichment. An agent given a list of company names from your CRM can research and append additional information — company size, recent permit activity, ownership structure, additional contact information — that makes your outreach more targeted and your account conversations more informed.

The Mindset Shift That Makes This Work

The operators who get the most from agentic AI are the ones who stop thinking about it as a tool and start thinking about it as a capability to deploy strategically.

A tool does what you tell it when you use it. A capability runs whether you're engaged with it or not. The Mac Mini running research tasks overnight while you're at home is a capability. The lead list in your CRM inbox when you sit down Monday morning is the output of a capability that ran without you.

Building that capability requires investment upfront — the setup time, the prompt development, the quality review of early outputs to refine the agent's instructions. An agent's first output on a new research task is rarely exactly what you want. You review it, identify what needs to change, update the prompt, and run it again. After two or three iterations, the output quality is consistent and the process is reliable.

That iteration cycle is the work. Once it's done, the capability runs indefinitely with minimal maintenance.

The other mindset shift is about where your time belongs. An hour you spend building a research agent that will run this task every week for the next two years is not an hour spent on research — it's an hour spent creating a perpetual research function. The ROI on that hour compounds in a way that no single hour of manual research ever could.

Start With One Use Case

The path to having a functional digital employee isn't to build everything at once. It's to identify one high-value repetitive research or data task that currently costs you significant time and automate that first.

For most service businesses the right starting point is lead list building for one specific target customer type. Pick the commercial customer category that generates the most recurring revenue for your business — contractors, property managers, or whichever segment is most valuable. Define exactly what a qualified prospect in that category looks like. Write a detailed research prompt. Run it. Review the output. Refine the prompt. Run it again.

When that process is reliable and the output quality is consistent, it becomes a recurring task you schedule rather than a project you manage. Then you identify the next high-value task and build the next capability.

Within six months of starting, an operator who approaches this systematically will have multiple automated research and intelligence functions running continuously — a digital employee that never calls out sick, never needs a raise, and gets better at its job every time you refine its instructions.

The Competitive Window

Here's the honest reality about where this capability sits right now.

Most service business owners don't know Claude Code exists. Most of the ones who do know it exists don't know how to deploy it for operational use. The operators building agentic AI workflows into their business development now are establishing an advantage that compounds with every month of iteration. Early adopters in every previous technology cycle — Google Ads, review generation, online booking — built leads that late adopters never fully closed.

That window won't stay open indefinitely. As these tools become more mainstream, more operators will adopt them and the advantage will compress. The operators who build the capability now — while setup requires some effort and most competitors haven't started — will have systems, processes, and refined prompts that took time to develop. Late adopters will be starting from scratch against operators who have two years of iteration behind them.

> *Your competitor is still doing this manually. Build the system that does it while you sleep.*

Go Deeper

For more on this and everything else in the book, the Haulers' Edge Newsletter goes deeper every week. Scan the code or subscribe free at HaulingHubb.com.

CH. 34 THE FUTURE OF BEING FOUND

"I skate to where the puck is going to be, not where it has been." — Wayne Gretzky

Something fundamental changed in how customers find service businesses and most haven't adjusted for it yet.

If you've paid an SEO company $500 to $1,500 a month and struggled to measure what you got for it, you're not alone. Most SEO work done for small service businesses in the past decade was built around a model that's actively changing — rankings, clicks, traffic metrics that didn't connect clearly to booked jobs. What this chapter describes is different in a specific way: it's built around the signals that determine whether your business gets named in AI-generated answers, voice search results, and zero-click local results — not just whether your website ranks. The tactics overlap with traditional SEO in places, but the goal is different. You're not trying to rank a page. You're trying to be the answer the system gives.

You built your website. You invested in SEO. Maybe you wrote blog posts. All of it was built on the assumption that a customer would search, see your listing, click through, and call. That model is breaking down faster than most people realize.

A growing majority of local searches now end without a click to any website — the customer gets their answer directly on the results page— from an AI-generated summary, a featured snippet, a Google Business Profile, or a voice assistant reading one result aloud. Your carefully built website may never enter the picture.

This isn't a future threat. It's the current reality. And the operators who keep optimizing only for clicks are playing a game that's actively changing underneath them.

From Traditional SEO to AI Search Visibility SEO

Traditional SEO had a clear goal: rank high, get clicks, bring visitors to your site. Success was measured in traffic. The whole strategy assumed the customer would travel from the search results page to your website before making a decision.

The reality is that the customer's journey increasingly ends on the results page itself. The question is no longer just "can they find my website" but "what do they see and hear about my business before they ever click anything."

That includes your Google Business Profile showing in the local map pack. Your review rating visible at a glance. An AI-generated overview that names you — or your competitor — as the answer. A voice assistant reading one business name when someone asks for a recommendation.

In all of these scenarios, your business is either present or it isn't. And being present in these zero-click moments often matters more than your website ranking because it's where the decision happens.

The businesses built to win in this environment have three things working for them: a strong, consistent review profile that AI systems treat as their primary trust signal, complete and accurate business listings across every platform that feeds search data, and website content structured so AI can extract and quote it directly. Traditional SEO is not dead — it feeds into AI Search Visibility SEO rather than competing with it. The fundamentals compound.

The Platforms Pulling Your Customers Away From You

Search used to mean Google, with a small percentage going to Bing. Today customers are finding local service businesses through an expanding set of AI-powered systems, each pulling from different data sources, each potentially naming your competitor instead of you.

Google AI Overviews — Google now puts AI-generated answers at the top of search results before anything else. They pull from your reviews, your business profile, and your website content. If your business isn't represented in that answer, the customer may never scroll far enough to find you. Your Google Business Profile data now feeds directly into these AI-generated answers, making structured profile information critical for local visibility *(digitalapplied.com, 2026)*. When a user searches "junk removal near me," the AI Overview may name specific businesses — pulling from review data, GBP content, and website structure — before the traditional results even appear. If your business is named in that answer, the customer may call without ever clicking anything. If it isn't, they may never scroll far enough to find you.

As covered in Chapter 13, local service queries are among the least affected by AI Overviews compared to informational content. But that's changing. AI-powered local packs are now rolling out on mobile, appearing on approximately 7% of tracked local keywords as of early 2026, with call buttons being replaced by images in several service industries *(Sterling Sky, January 2026)*. The format of local search results is actively shifting and the operators who understand it now will have the adjustment behind them when it becomes mainstream.

ChatGPT, Perplexity, and AI search tools are replacing traditional search for a growing segment of customers — particularly younger demographics and professional buyers. When someone asks ChatGPT "what's the best junk removal service near me," the AI assembles an answer from web content, reviews, and business data. AI tools like these heavily reference comparative content — securing mentions in "best local providers" roundups and review aggregators drives citations from these platforms *(webpronews.com, 2026)*. Your business being named in these answers is a new form of visibility that doesn't exist in traditional SEO frameworks but is increasingly driving real inbound calls.

Voice search remains the most demanding zero-click scenario. A voice query produces one answer, spoken aloud, with no screen to scroll and no list to browse. Each assistant pulls from different sources: Google Assistant relies primarily on your Google Business Profile and local pack ranking. Apple's Siri pulls from Apple Maps and Yelp — if your Apple Maps listing is incomplete or your Yelp reviews are thin, Siri may ignore your business entirely. Amazon's Alexa draws heavily on Yelp for local business queries. These aren't marginal channels — a significant percentage of service queries, especially "near me" searches, are being handled by voice assistants on phones, smart speakers, and in cars.

The unifying principle across all of these: you have to win the recommendation, not just the ranking. It's not enough to be on page one. You need to be the answer that page one — or the voice assistant — delivers.

Search algorithms change regularly. The principles in this section — review velocity, GBP optimization, citation consistency, structured content — have been directionally stable for years. Specific ranking factors and their relative weight will evolve. For the most current local SEO guidance, the newsletter covers what's changing in real time at HaulingHubb.com.

What AI Systems See When They Look at Your Business

When an AI assembles an answer about local service businesses, it pulls from three primary data sources. Understanding these sources tells you exactly where to invest your effort.

Reviews — quantity, velocity, and engagement. Reviews are the single most important factor for local AI visibility. When an AI synthesizes "the best" or "top-rated" in a category, it's referencing review data — count, average rating, recency, and the specific language customers use. A competitor with 300 reviews averaging 4.8 stars will be named over your business with 40 reviews at 4.3, regardless of how well your website is built.

Consistent weekly review velocity now matters more than total volume. Businesses maintaining a steady weekly review cadence show stronger map pack stability than companies relying on large but stagnant review totals — fresh engagement signals outweigh historical volume *(local SEO research, 2026)*. Two to three new reviews every week, consistently, is more valuable than a burst campaign that produces fifty reviews in a month and then goes quiet.

Responding to reviews matters as much as generating them. Prompt responses to every review build engagement signals that Google weighs in local ranking — review quality, recency, and the keywords customers use all factor into how AI systems evaluate and recommend your business *(kexworks.com, 2026)*. A business that responds to every review signals active management. An inactive review profile signals a business that may not be current, which is exactly the wrong signal to send.

Encourage customers to include specifics in their reviews — the service they received, the city where the job happened. "Fast garage cleanout in Tampa" in review text provides AI systems with keyword context they use when assembling location-specific answers.

Business profiles and listing consistency. When an AI needs factual details about your business — address, phone, hours, services, service area — it pulls from structured listings. Your

Google Business Profile is the primary source, but Yelp, Facebook, Bing Places, Apple Maps, Angi, and industry-specific directories all feed into the data picture search engines build about you.

The critical word is consistency. Your Name, Address, and Phone number — NAP — must be identical everywhere. Not similar. Identical. If your NAP varies anywhere online, your visibility will be reduced regardless of keyword optimization. An old address on one directory, a different phone format on another, a slightly different business name on a third — all of these create ambiguity that AI systems resolve by reducing confidence in your business as a local entity.

Incomplete profiles are as damaging as inconsistent ones. If your Google listing doesn't include a service you offer, an AI may not know you provide it and exclude you from answers about that service. If your hours are wrong, a voice assistant may tell a caller you're closed when you're open. Your profiles are your business's data source for every AI system that might recommend you. Treat them accordingly.

Structured content and schema. AI systems are only as useful as the data they can extract. They crawl websites for content, but they especially rely on structured data — schema markup, FAQ sections, and clearly formatted content that answers specific questions directly.

Schema markup tells Google exactly what your business does in a format it can read directly. It doesn't guarantee visibility but it removes ambiguity — Google has to guess less about who you are and what you offer. Ask your web person to set it up. It takes an hour. FAQ schema makes your question-and-answer content eligible to appear directly in search results as expandable snippets. Service schema gives AI systems specific information about each service you offer.

Make Your Business AI-Quotable

The goal of AI Search Visibility SEO is to ensure that when an AI assembles an answer about services in your market, your business has the strongest possible data signal. Here's how to build that signal across each dimension.

Reviews: treat them like a core business metric.

Set a target and track it. Getting to 200 or more Google reviews in your market is achievable for an active service business and provides a meaningful competitive moat. Triple-digit review counts get cited in AI summaries — "known for 150+ five-star reviews" is the kind of language that appears directly in AI-generated answers. Below that threshold, you're often invisible to the AI comparison.

Build the review generation system covered in Chapter 10. Two to three new reviews per week is the velocity that signals consistent activity. Service Hubb AI handles this automatically once configured.

Respond to every review within 48 hours. Thank positive reviewers specifically. Address negative reviews professionally and briefly. Every response adds to the engagement signal that AI systems use to determine whether your business is actively managed.

Google Business Profile: your most important digital asset.

According to Advice Local's 2026 research, GBP is now the top ranking factor for Local Pack and Maps, and inactive profiles are increasingly penalized with lower visibility *(webpronews.com, 2026)*. Full optimization means every field complete, every section filled, every feature enabled.

Write your business description using natural language your customers actually search — specific services, specific service area, specific differentiators. List every service you offer with short descriptions. Fill in all relevant attributes — locally owned, free estimates, online booking available. Every completed field is a data point the AI can use when assembling a recommendation.

Google now treats your GBP as a real-time discovery hub, not a static listing *(digitalapplied.com, 2026)*. Post updates at least once per week — a before-and-after from a job, a seasonal reminder, a tip relevant to your customer base. Posting frequency is an engagement signal that feeds the algorithm.

Upload photos regularly and make them work harder. Geotagged photos now verify physical presence in your service area — Google uses image metadata and AI recognition to confirm that work is actually happening in the locations you claim to serve. Enable GPS on your phone when taking job site photos and upload them directly. This is a newer ranking signal most operators aren't using yet.

Preload the Q&A section with your real FAQs and answer them from your business account. This content is indexable, surfaces in voice search results, and gives AI systems direct question-and-answer content to quote. Enable direct booking integration if Google offers it for your business type — booking directly from GBP creates behavioral engagement signals that feed local ranking *(digitalapplied.com, 2026)*.

NAP consistency across all platforms.

Pick one exact format for your business name, address, and phone number and replicate it identically everywhere your business appears. Check the primary platforms: Yelp, Bing Places, Apple Maps, Facebook, Angi, HomeAdvisor, Thumbtack, the Better Business Bureau, your local Chamber of Commerce, and any industry-specific directories. Use a tool like Moz Local or Yext to audit your citations and identify inconsistencies, then fix them on the most important platforms manually.

Bing Places imports directly from your GBP — set it up once. Apple Maps is essential for Siri — claim your listing via Apple Maps Connect and verify your information is current. Yelp is critical for Alexa and Siri voice recommendations — maintain your listing even if you're primarily focused on Google.

Website content built to be extracted.

Write answers, not just articles. Write short direct answers. AI pulls from them. Long paragraphs get skipped.

The content that AI search pulls from most consistently isn't blog posts or general service descriptions. It's direct honest answers to the questions buyers are actually researching. Cost and pricing. Problems and negatives about the service. Comparisons between providers. Reviews and validation. Best-of content for the local market.

Most service businesses avoid publishing this content because it feels uncomfortable. That's the competitive opportunity. When a buyer asks an AI assistant what junk removal costs in your city or what the downsides of hiring a hauling company are — the business that has published a clear honest answer to that question is the one that gets cited. The one that says “call for a quote” is invisible.

We covered building these pages in Chapter 13. In the context of AI search they're not just good website content — they're the specific signal that tells AI systems your business is a credible, transparent, and authoritative source worth recommending. Build them once and they work across both traditional search and AI-generated results simultaneously.

A FAQ section with specific questions and concise two to three sentence answers is the highest-value content format for AI visibility. “How much does junk removal cost in [city]?” followed by a specific answer that includes a price range, what affects the price, and what's included — that's the format AI systems quote directly. “Learn more on our pricing page” is the format they skip.

Use natural conversational language. AI systems are trained on how humans speak and search, and they prefer content that mirrors that. Instead of “We are the premier provider of residential and commercial waste removal services in the Tampa Bay metropolitan area,” write “We provide fast junk removal and dumpster rentals for homeowners and contractors across Tampa Bay.” Same information, dramatically more quotable.

Each service area page needs genuinely unique content specific to that location — local context, community involvement, neighborhood references. Generic pages that only swap city names are now penalized by Google's algorithm *(digitalapplied.com, 2026)*. “Junk Removal in Tampa” can't be a copy of “Junk Removal in St. Petersburg” with the city name changed. Each page needs content that reflects the actual character of that service area.

Traditional Local SEO Still Pays

With all of this, you might wonder whether the classic local SEO work still matters. It does — more than ever.

The actions that improve your map pack ranking — reviews, GBP completeness, schema markup, NAP consistency, service area pages — are the same actions that make your business AI-quotable. You're not optimizing for two separate systems. You're building one foundation that serves both. The difference from traditional SEO isn't the tactics. It's the goal: you're not trying to rank a page. You're trying to be the answer the system gives.

The critical distinction is in expectations. Even well-ranked businesses are seeing fewer calls from GBP listings as Google changes how local results display on mobile, replacing call buttons with images in several industries *(Sterling Sky, January 2026)*. This doesn't mean stop investing in local SEO — it means measure success through total customer contacts, not just website clicks. Calls, messages, direction requests, and direct bookings from GBP are all signals that the strategy is working, even if website traffic is flat or declining.

Clients running Google Ads alongside strong local SEO are seeing better ROI than previously because ads now carry features — including call buttons — that organic listings are losing *(Sterling Sky, January 2026)*. The paid and organic strategies reinforce each other. Organic builds the trust and entity reputation that makes your ads convert better. Paid amplifies visibility when organic alone isn't enough.

The businesses that invest in both will dominate both the map pack and the AI-generated answers that are increasingly where customer decisions happen.

Your Six-Month AI Search Visibility Plan

Month 1: GBP audit and enrichment. Log into your GBP dashboard and fill in every field. Write a clear, specific business description using natural language. Upload at least twenty real photos — with GPS enabled for geo-tagging. Set up a weekly posting cadence. Preload the Q&A section with five to seven real FAQs and answer them from your business account. Enable every applicable feature — messaging, booking links, service listings.

Months 1–6, ongoing: review generation system. Two to three new reviews per week is the target. Automate the follow-up request through a CRM so it runs without manual management. Respond to every review within 48 hours. By month six you should have meaningfully more reviews than you started with, consistent recent dates, and an active response pattern that signals an engaged business.

Months 2–3: service area content. Build out or revamp your service area pages — one per major city or town you serve. Each page needs genuinely unique content specific to that location: local context, neighborhood references, any community involvement in that area. Add FAQ sections with schema markup to your primary service pages. Write short, answer-first content that directly responds to the questions you get asked most often.

Month 4: citation audit and cleanup. Systematically audit your NAP consistency across every platform your business appears on. Fix every inconsistency. Submit to any quality directories you're missing. Verify your Bing Places and Apple Maps listings specifically — these feed Microsoft Copilot and Siri respectively and are frequently overlooked.

Months 4–5: community signals and backlinks. Sponsor a local event and get mentioned on their website. Partner with a charity and document it on your GBP and your site. Join your local Chamber of Commerce and ensure the directory listing is current. Getting your name mentioned on local websites tells Google you're real and established in the community. That trust compounds over time.

Month 6 and ongoing: measurement and iteration. Review your GBP Insights — calls, direction requests, profile views. Check Google Search Console for impressions versus clicks — impressions rising while clicks hold flat is evidence of AI Search Visibility SEO working. Track your review count and velocity. Identify the geography or service where you're not showing up and diagnose why — usually it's a missing service area page, inconsistent NAP, or insufficient review volume for that specific location.

AI Search Visibility SEO isn't a project with an end date. It's an ongoing maintenance discipline, like keeping your trucks running. The businesses that treat it that way will compound their advantage over the ones that do it once and move on.

Whether the customer finds you through a clicked link, an AI summary, a voice assistant, or a direct GBP call — you want to be the answer they get. Build the data foundation that makes that happen consistently, and the customers come to you whether they clicked or not.

> *The future of being found isn't on your website. It's everywhere your business data lives. Feed it well.*

Go Deeper

For more on this and everything else in the book, the Haulers' Edge Newsletter goes deeper every week. Scan the code or subscribe free at HaulingHubb.com.

CH. 35 YOUR NEXT COMPETITOR USED TO WEAR A TIE

"Every battle is won before it is fought." — *Sun Tzu*

There's a new competitor entering your market. You probably haven't noticed them yet. They're not coming from another hauling company or a competing franchise. They're not someone who grew up in the trades or spent years learning the business from the ground up.

They used to work in marketing. Or finance. Or project management. Or tech. They got laid off, or burned out, or decided the corporate ladder wasn't worth climbing anymore. And they looked at your industry — at the margins, at the demand, at the lack of sophistication in how most operators run their businesses — and they saw an opportunity.

They're not starting from zero. They're starting with skills that most trade owners spent years either not developing or actively avoiding. They know how to run Google Ads because they managed paid search campaigns. They understand financial modeling because they built spreadsheets for a living. They know how to write copy, build a brand, hire strategically, and use AI tools that most operators in your market have never heard of.

And they have nothing to unlearn.

That's the competitive threat nobody in the service industry is talking about seriously enough. This chapter is about understanding it clearly — and about what you do to stay ahead of it.

The Migration Is Already Happening

The displacement of white-collar workers by automation and

AI is not a future event. It's been accelerating since 2020 and the pace is increasing. The people affected are not low-skill workers. They're marketers, analysts, coordinators, project managers, customer success professionals, recruiters, and mid-level managers whose jobs got absorbed by software, AI tools, and organizational restructuring.

BLS data on layoffs, job openings, and labor market flows — combined with widely reported corporate restructuring announcements — points to white-collar job displacement during 2024–2025 at levels not seen since the 2008 financial crisis, with management, administrative, and professional service roles taking the steepest cuts. These aren't people who will take another white-collar job — the market for those roles is increasingly competitive and AI is continuing to compress it. Many of them are looking at entrepreneurship as the alternative.

And when they look at the service business sector — junk removal, landscaping, cleaning, HVAC, pest control, contracting — they see what you see every day from the inside, but with fresh eyes that notice something different. They see an industry with strong, consistent demand, relatively low startup capital requirements, and operators who are almost universally unsophisticated in the areas where these newcomers have deep expertise.

They see an open door.

What They're Bringing That You Don't Have

The white-collar migrant entering your market isn't stronger than you physically. They may not know how to run a job as efficiently as you do. They probably don't have your customer relationships or your ten years of operational knowledge.

But they have things that are increasingly more valuable than those traditional advantages.

Marketing fluency. A former digital marketing manager knows how to build a Google Ads campaign that converts, how to structure a landing page for maximum conversion rate, how to use retargeting to stay in front of prospects who visited the website but didn't book. They don't need to learn this the hard way over three years of wasted ad spend. They know it coming in. They'll outperform your Google Ads from month one — not because they worked harder but because they started with competence you had to earn.

AI tool proficiency. This is the one that matters most right now. The white-collar professional who was displaced partly because AI tools replaced their role is, by definition, someone who understands how to use AI tools. They know how to prompt Claude or ChatGPT to produce marketing copy, write job descriptions, build competitive research, draft sales scripts, and generate content at scale. They know how to build automation sequences in a CRM. They know what Claude Code is and how to deploy it to build lead lists — exactly what we covered in Chapter 33. They arrive in your market with an AI-enabled operational capacity that most legacy operators are still figuring out exists.

Financial sophistication. The former analyst or finance professional entering the trades understands profit margins, cash flow management, job costing, and business valuation in ways that take most trade owners years of trial and error to understand. They won't underprice jobs for years before figuring out their true cost structure. They'll build a P&L from day one. They'll track job-level margins. They'll manage their cash flow with the same rigor they applied in corporate.

Systems thinking. Corporate careers teach people to build processes, document workflows, and create scalable systems.

The person who spent five years managing cross-functional projects at a marketing agency understands SOPs, training documentation, and delegation. They'll build a business that doesn't depend entirely on them faster than most who learned the trades from scratch.

Brand building. The former brand manager or content strategist who starts a cleaning company will build a visual identity, a consistent online presence, and a content strategy in the first sixty days that most established competitors in their market don't have after ten years. Professional presentation — website, branding, photography, social media — is table stakes knowledge for these people, not an afterthought.

The Threat Model

Here's what this looks like in practice.

A former corporate project manager in your market decides to start a junk removal company. In their first month, they set up a properly structured Google Ads campaign targeting high-intent local keywords. They build a website that converts because they understand the principles from their marketing experience. They configure a CRM and set up automated follow-up sequences before they take their first call. They deploy an AI voice agent because they read about it and understood it immediately. They build a lead research agent in Claude Code over a weekend because they have the technical comfort that most operators don't.

By month three, they're outperforming operators who have been in the market for five years on every digital touchpoint — response time, conversion rate, review velocity, GBP completeness. Their pricing is correct from the start because they modeled their cost structure before they launched. Their customer experience is consistent because they documented it before they hired anyone.

They are, in most measurable ways, running a more sophisticated operation than most established competitors in the market — before they've learned the physical and operational nuances that you know cold.

And here's the part that should concern you most: they're going to learn those operational nuances. Fast. Because they're smart, they're motivated, and they have good systems for learning. In twelve to eighteen months, they'll have the operational competence to match yours and the marketing and systems sophistication to outperform you.

That's your competitive window. It's not infinite.

Why Legacy Operators Are Vulnerable

The operators most at risk from the white-collar migrant competitor are the ones who built their businesses on advantages that no longer differentiate.

Answering the phone consistently used to be an edge. Today an AI voice agent answers every call. Being reachable used to win jobs. Today automated follow-up sequences run around the clock. Having a website used to set you apart. Today having a website that converts requires the kind of optimization knowledge that white-collar migrants arrive with.

The advantages that remain meaningful are the ones that take genuine time to build and can't be immediately replicated by someone with business sophistication but no operational history. Your customer relationships with contractors and property managers who have trusted you for years. Your knowledge of your specific market's quirks — the transfer stations, the permit requirements, the neighborhoods, the seasonal patterns. Your reputation in the community, built through years of showing up. Your team, if you've built one that operates at a high level.

These are real advantages. But they're not permanent if you're standing still while a better-equipped competitor is building operational sophistication alongside their own version of those advantages.

The operators who are most protected are the ones who've already built what this book describes — strong review profiles, documented systems, AI-enabled operations, recurring commercial relationships, financial clarity. Not because of this specific threat but because building those things makes your business better by every measure. The white-collar migrant threat makes the urgency more visible, but the prescription is the same.

What You Do About It

The answer isn't to become a white-collar worker. It's to adopt the capabilities they bring while keeping the advantages they'll spend years trying to acquire.

Close the marketing gap now. If your Google Ads aren't optimized, your website isn't converting, your follow-up sequences aren't running, and your review generation isn't systematic — fix those things before a competitor who arrives with those skills out-of-the-box takes market share you didn't know was at risk. Chapters 13 and 14 give you the marketing framework; Chapters 16 and 17 give you the conversion and follow-up process. The tools are available. The window to build these systems before sophisticated competition arrives is narrowing.

Build the AI capability before it's table stakes. The operators who are deploying AI voice agents, follow-up automation, routing optimization, and Claude Code research agents now are building an advantage that compounds. These capabilities will eventually be widespread enough that having them doesn't differentiate — not having them just means you're behind. The only question is how much lead time you build before that happens. Get there while being ahead still means something.

Lock in your recurring relationships. The commercial clients who call you first — the contractors, property managers, real estate investors — are your most durable competitive moat. These relationships took years to build and can't be replicated quickly by a newcomer regardless of how sophisticated their marketing is. A property manager who has worked with you for three years and trusts your crew isn't switching to a new entrant with a better website. Deepen and formalize these relationships as covered in Chapter 27.

Build the brand that the community knows. Yard signs, truck branding, community sponsorships, consistent presence in the neighborhoods you serve — this is local recognition that compounds over years and can't be bought by someone who just entered the market. The new competitor can run better ads than you in month one. They can't replace ten years of your name showing up on trucks in the neighborhood.

Hire and develop before you need to. A business with a functioning team, documented SOPs, and a hiring process is harder to compete with than a solo operator or owner-dependent operation. The white-collar migrant who launches solo will eventually need to build a team. If you've already built yours and systematized it, you have an organizational advantage that compounds with every hire they make badly.

One action this week that's specific to this threat: search your core service term on Google right now. Look at the first page of results. Count how many operator listings have professional branding, a clean website, published pricing, and a review profile above 100. If two or three results look like they were built by someone with a marketing background rather than someone who grew up in the trades — the migration has already reached your market. That's your baseline. Everything in "What You Do About It" above is the response.

Why "They Don't Know the Work" Isn't the Defense You Think It Is

The most common response from experienced operators reading this chapter is: "These people don't know the work. They'll fail." It's a reasonable instinct and it's partially right. Operational knowledge — how to run a job efficiently, manage a crew on-site, handle a difficult customer face-to-face, navigate the physical and logistical reality of the trade — takes time to build and can't be faked. A new entrant with a marketing background doesn't have it on day one.

Here's why it's not the defense it feels like. They don't need to have it. They need to hire someone who does — and they will, within the first month, because they understand that operational execution is a solvable problem. A former marketing director who launches a junk removal company doesn't try to figure out how to run a job alone. They hire an experienced crew lead from their second week, pay them well because they understand labor economics, and focus their own energy on the customer acquisition and systems work where they have genuine advantage. The operational knowledge gap gets closed by hiring. The marketing, financial, and technology gaps that most legacy operators have — those don't close as quickly.

The moat that actually protects you isn't knowing the work. It's the years of trusted relationships, community reputation, and commercial contracts that new money can't buy overnight. That's the real defense. Build it before you need it.

The Opportunity in the Threat

Here's the reframe worth sitting with.

The same forces that are sending white-collar migrants into your market are creating the tools that give you — the operator who already has the market knowledge, the customer relationships, the operational experience, and the reputation — the ability to run like a sophisticated business without needing to have a corporate background.

Claude Code exists. Service Hubb AI exists. Route optimization software exists. Automated follow-up sequences exist. AI-assisted quoting exists. The AI Search Visibility SEO framework in the previous chapter exists.

Every capability the white-collar migrant is bringing into your market is available to you. The difference is that you already have everything they have to spend years earning. You have the market knowledge. You have the relationships. You have the reputation. You know what a ton of construction debris actually weighs and what the transfer station charges for it on a Monday morning in January.

They have to learn what you already know. You just have to learn what they already know. That asymmetry is in your favor if you act on it.

The operators who read this chapter and move on without changing anything are the ones who will look back and remember when the competition was easier. The ones who read it and build accordingly will be the ones that the next white-collar migrant looks at when they enter the market and says "that operation is too far ahead — I need a different market."

Be the one they don't want to compete with.

> *The threat is real. The advantage is still yours. But only if you build it before they do.*

Go Deeper

For more on this and everything else in the book, the Haulers' Edge Newsletter goes deeper every week. Scan the code or subscribe free at HaulingHubb.com.

CH. 36 HISTORY RHYMES

> *"Those who cannot remember the past are condemned to repeat it." — George Santayana*

History doesn't repeat itself. It rhymes.

If you've been in business long enough, you know the rhyme. A new technology shows up. The early adopters jump in and start pulling ahead. Everyone else shrugs it off and keeps doing things the old way. Then one day the new way becomes the only way — and the ones who ignored it get steamrolled.

That rhyme has played out with every major leap forward in the modern economy. The mechanized loom. The personal computer. The internet. Each time, business owners were given the same choice: adapt early and ride the wave, or resist and get left behind.

The pattern is accelerating. Each technology cycle has moved faster than the one before it. AI is moving faster than any previous cycle. If you're telling yourself "I'll wait and see how this plays out" — you're already behind.

The Pattern That Keeps Repeating

Every major technology shift in the last two hundred years has followed the same arc. Early adopters use it while everyone else ignores it. Then it becomes essential. Then it becomes the replacement for anyone who waited too long.

The mechanized loom wiped out skilled weavers who'd spent decades mastering their craft. Not because handmade cloth wasn't good — it was — but because a machine could produce in a day what dozens of craftsmen produced in a week. The economics made the argument. The Luddites smashed machines and it didn't matter. The loom was faster, cheaper, and scalable, and that was the whole story.

The personal computer did the same thing to office work. Typing pools. Accounting teams balancing ledgers by hand. Filing clerks. Switchboard operators. One worker with Excel outproduced entire teams. The companies that embraced it early gained speed and cut costs. The ones that held onto typewriters disappeared.

The internet eliminated entire profit centers that felt permanent. Blockbuster. Travel agents. Newspaper classifieds. Not because the old model was bad — it worked well until it didn't. But once the new option existed, customers shifted, and the businesses that ignored it long enough couldn't recover.

AI is following the same pattern, and doing it faster than any previous cycle. The adoption curve is steeper. The window to act is shorter.

Why AI Is Different This Time

Every previous revolution disrupted lives, rewrote industries, and left winners and losers. AI is different in three specific ways that matter for a service business owner.

It's hitting white-collar and blue-collar simultaneously. The Industrial Revolution started with blue-collar trades. The Digital Revolution wiped out clerical and administrative work. AI is eating both at once. It drafts contracts, writes ad copy, analyzes financials, and generates creative work. It also diagnoses equipment problems and operates machinery with robotic precision. That's not someday — that's now. AI is already working through middle-class roles that once felt untouchable, while simultaneously automating physical and operational work in the trades.

It scales instantly. Factories took years to build. Offices took months to set up. Hiring and training took weeks. AI is code. Once built, it can be deployed at scale overnight. That means the early adopter advantage window is shorter than most people expect. The gap between the operator who adopts early and the one who waits is closing faster than any previous technology cycle.

It's already in the tools you use. The loom required new factories. The PC required new hardware. The internet required websites and servers. AI is being built directly into the software you already have — your CRM, your scheduling platform, your accounting software, your ad platform. One day you're using the old version. The next day there's a feature that can draft, analyze, automate, and optimize in seconds. Adoption doesn't require buying something new. It just shows up in what you already have.

There's a fourth dimension worth naming that cuts across all three of the above. Every disruption in the examples above had a technology component — but the technology was rarely the whole story. Netflix didn't win because streaming was technically superior to DVDs. They won because they were willing to let customers watch whenever they wanted, without late fees, without a trip to the store. Expedia didn't win because websites were better than phone calls. They won because they were willing to show all the options including competitors. The disruption was always partly about willingness — the new entrant's willingness to do what the established player refused to do because it was uncomfortable or threatened the existing model. The new competitor entering your service market isn't just bringing better marketing tools. They may be willing to publish their pricing when you won't, show their full team when you don't, answer every question online when you haven't, and be completely transparent about their process when you've always kept it vague. That willingness is as disruptive as any technology they bring with them. The antidote isn't better technology. It's being more willing than they are — before they arrive.

The Human Reaction Curve

Most operators go through the same emotional cycle every time a major technology shows up. First, denial — the belief that this one won't affect their business. The weavers thought people would always prefer handmade cloth. Today that sounds like "AI is fine for writing blog posts but my business is relationship-based" or "my customers just call me." That's what taxi companies said before Uber.

Then anger — the frustration that this feels unfair. The Luddites smashed looms. Print unions fought automation. Today that sounds like "AI is stealing jobs" or "these new guys are undercutting everyone." The anger might feel justified. It doesn't slow adoption.

Then bargaining — dabbling at the edges without committing. Testing AI for one thing while leaving everything else untouched. It feels safe. But dabbling doesn't keep pace with competitors who go all in.

Then acceptance — and the ones who get here early enough to act on it are the ones who come out ahead. That looks like baking AI into how the business actually runs: quoting, scheduling, follow-ups, customer communication, lead generation. That's not a someday project. It's the only stage where decisive action still produces a competitive advantage. Wait long enough to get here and the early movers have already locked down the best ground.

Wherever you find yourself on this curve, the next move is specific. If you haven't used any AI tool in your business yet, open one today and ask it to draft a follow-up text for a recent estimate — that's the entry point. If you've used AI occasionally but inconsistently, pick one function from Ch. 31 and run it every day for two weeks until it's habit. If you're already integrating AI across your operation, Ch. 33 is your next chapter — building a research agent creates the kind of compounding operational advantage that separates the early movers from everyone else.

The objection forming right now is probably this: "My business isn't Blockbuster. I provide a physical, local service. Digital disruption hit information businesses. Mine is different."

It's a fair objection and it's worth answering directly. You're right that the service itself — hauling, cleaning, landscaping, HVAC — can't be digitized. The work is physical and it always will be. But the disruption in the historical examples above wasn't primarily about the product. It was about customer acquisition, communication, and operational efficiency — all of which are digital, and all of which are exactly where AI-enabled competitors have an advantage over legacy operators who haven't modernized. Netflix didn't replace the experience of watching a movie. It replaced the process of finding and accessing one. The new competitor entering your market isn't trying to replace your crew. They're replacing how customers find you, evaluate you, and decide to book you. That layer is entirely digital — and that's where the disruption lands.

The Coming Flood

Here's the part most service business owners don't see coming. AI won't just change how you operate. It will change who you compete against.

When middle-class jobs get absorbed by AI, a meaningful percentage of those displaced workers won't spend years retraining for new white-collar careers. They'll look for businesses they can start with relatively low capital, with immediate demand, and without years of technical training. They'll land in service trades.

That means more cleaning companies, more landscapers, more pressure washers, more movers — and more operators in every service trade.

We've seen this pattern before. After the 2008 financial crisis, residential cleaning saw a flood of new operators — many displaced white-collar workers who decided to try a service business. Within two years, established cleaning companies were fighting rising lead costs, more competition for the same customers, and the need to either specialize or scale just to protect margins.

Between 2015 and 2018, pressure washing businesses roughly doubled in most markets based on our observation of local search results and Google Maps listings across multiple regions. Many new owners weren't tradesmen — they were laid-off sales reps, office workers, and managers looking for income. Google Ads cost per click more than doubled. Undercutting became rampant. Customers became conditioned to shop around.

During COVID in 2020 and 2021, two forces collided in junk removal. Demand spiked as people stuck at home cleaned out everything. Supply spiked as laid-off workers launched hauling businesses. In the markets we tracked directly — including our own — the number of junk removal companies on Google Maps doubled in under two years. Ad costs rose. Price undercutting intensified. Customer loyalty was tested with more choices than ever before. The same dynamic played out in pressure washing, lawn care, and cleaning — any service trade with low startup costs and visible demand attracted a surge of new entrants.

That COVID surge eventually tapered as the unusual demand conditions normalized. If the pattern holds — and every previous technology displacement cycle suggests it will — the AI-driven displacement wave will be different. It won't be temporary. The people entering these markets won't be waiting out a disruption — they'll be building livelihoods. And as covered in Chapter 35, the new entrant this time won't just be someone who needs quick income. Many will be former marketers, analysts, and project managers who arrive in your market with capabilities most established operators don't have.

How fast this plays out will vary by market, by trade, and by region. Some service markets are already feeling the pressure. Others have more runway. The honest answer is that nobody can tell you exactly when the wave reaches your specific zip code. What history tells you — clearly and consistently — is that the operators who built before the pressure arrived kept their position, and the ones who waited until they felt it were already behind.

Every low-barrier service industry eventually hits the same wall: too many operators chasing the same pool of customers. The pattern is brutal but predictable.

In Adimize client data tracking local service categories, we've seen Google Ads cost per click for "pest control" jump from roughly $8 to $18 over a three-year period in competitive markets as more operators entered and all went straight to paid search. Established companies had to double their ad spend just to maintain the same visibility. Those who didn't saw lead flow drop by 30–40% in a single year.

The moving industry has been saturated for decades. Every year, new operators jump in. The result is cutthroat underpricing, inconsistent reliability, and customers burned by fly-by-night operators. The survivors didn't win by being cheapest. They won by locking in commercial contracts and building brand trust strong enough that customers paid a premium to avoid the risk of the unknown.

The parallel to every other service trade is direct. Once enough new operators start competing for the same local keywords, ad costs rise, average job prices fall as undercutting becomes the default competitive move, and margins compress. This isn't speculation — it's the documented outcome of every previous saturation cycle in every service industry with low barriers to entry.

The question isn't whether this happens. It's whether you have your position built before it does. The defense against every dynamic described above is the same: build your position before the pressure arrives. Here's each layer of that defense.

The Six-Layer Moat

When markets flood, the operators who survive and thrive aren't necessarily the ones who've been in business longest. They're the ones who built the things that new entrants can't quickly replicate — regardless of how sophisticated those new entrants are. There are six layers. Stack all of them and the gap becomes nearly impossible to close.

Review dominance. Your review profile is your most durable trust signal. When a new competitor enters your market with better marketing than you but 8 reviews against your 250, the customer chooses you before the conversation even starts. Aim for 200 or more five-star reviews and build toward them systematically. The automated review generation process in a CRM makes this consistent rather than dependent on remembering to ask. Build this before the market crowds — review volume that took years to accumulate can't be bought or rushed.

Brand recognition. When your trucks are visible in every neighborhood you serve, when your name shows up consistently on job sites and in community events, when people recognize your colors before they read your name — that's brand equity a new entrant can't manufacture. Wrapped trucks are moving billboards. Uniform consistency signals professionalism. Community sponsorships create goodwill that advertising can't replicate. Build the local presence that makes your name the one that comes to mind when someone needs what you do.

The operators who survive saturation aren't always the ones who've been around longest or run the most ads. They're the ones who became the most known and most trusted name in their specific market. Known means your trucks are visible, your name comes up in conversations, and people recognize you before they search. Trusted means your reviews are consistent, your crew shows up the way you said they would, and the customers who've used you once don't hesitate to use you again. Those two things together — known and trusted — create a position that a new entrant with better marketing cannot quickly replicate. Marketing gets you found. Known and trusted gets you chosen without comparison.

AI Search Visibility SEO. The local search landscape is shifting from traditional SEO toward AI-generated answers and zero-click results. Building your digital presence now — optimized GBP, consistent NAP across all platforms, structured content that AI can extract and quote, geotagged photo uploads, consistent review velocity — creates a visibility foundation that compounds. New entrants who don't understand this framework will be invisible in the AI search landscape that increasingly determines which businesses get the call.

Commercial contracts. Residential jobs are transactional. A property manager who calls you for every unit turnover, a contractor who orders a dumpster every time a project starts — these relationships represent recurring revenue that no amount of competitor advertising can easily disrupt. The relationship, the track record, the established workflow — these have real switching costs that protect you even when a better-marketed competitor shows up. Formalize your best commercial relationships as covered in Chapter 27.

Community integration. Price shoppers are fickle. Community roots aren't. When your business sponsors the youth sports league, helps with the neighborhood cleanup, shows up with a truck at the charity event — those families remember you when the need arises. That kind of local embeddedness is genuinely difficult to replicate quickly. It accumulates over years of consistent showing up.

Operational efficiency. When competition intensifies and prices compress, the operator running the tightest operation maintains margin while competitors bleed. Standardized equipment, optimized routes, documented SOPs, automated workflows — all of these allow you to remain profitable at price points that disorganized competitors can't sustain. The inefficient operator gets squeezed out not because they were outcompeted on customer experience but because their cost structure couldn't survive the price pressure.

None of these moat layers is built overnight. Each one takes consistent investment over months and years. The competitive advantage comes from stacking all six — reviews plus brand plus visibility plus contracts plus community plus efficiency. That's the Six-Layer Moat. A new entrant, regardless of their background or sophistication, cannot replicate all six quickly.

There's one more advantage worth naming that the moat section doesn't fully capture. The physical, trust-based, relationship-driven nature of service work creates a natural protection that no amount of marketing sophistication can immediately overcome. A property manager who has trusted you for three years, who knows your crew shows up on time and doesn't create problems, isn't switching to a better-marketed newcomer easily. A contractor who has you baked into their workflow isn't testing an unknown vendor on a deadline-sensitive project. The switching cost of an established service relationship is real and it grows every month you keep delivering. Your moat accelerates that protection — but even without it, the nature of the work itself gives you more runway than operators in purely digital industries ever had when disruption arrived.

Building Your Moat — Where to Start

Here's the honest reality. The AI-driven displacement wave is already moving in some markets and approaching others. Nobody can tell you exactly when it reaches yours — but the pattern is consistent enough across every previous technology cycle that waiting for confirmation before acting is the same mistake the Blockbuster executives made in 2007. The operators who are already positioned when it arrives will be the ones who started building before they had to.

A focused period of consistent moat-building — starting now — is enough to establish a position that will hold through the saturation cycle. It doesn't have to be built all at once, but it has to be started. The foundation you build today is what compounds into something durable when the pressure arrives.

The first ninety days: generate two to three new reviews per week, fill out your Google Business Profile completely, make contact with five potential commercial partners, and document your core job workflow. These four actions cost nothing but time and they start building every layer of your moat simultaneously.

Six to twelve months: cross seventy-five five-star reviews, wrap at least one truck, build city-specific service pages for your primary markets, secure at least one recurring commercial relationship, and automate your follow-up and review request sequences.

Eighteen to twenty-four months: reach two hundred or more reviews, brand every truck in your fleet, lock in three or more recurring commercial contracts, and have your AI-enabled systems — voice agent, follow-up automation, routing optimization — running as standard operations.

By that point, the new competitor who enters your market will look at what you've built and make a calculation. Your review count, your brand visibility, your commercial relationships, your operational sophistication. And they'll either decide the gap is too wide to close or they'll spend years trying while you keep compounding.

That's not luck. That's the result of starting before the wave hits instead of scrambling after it does.

The Paddle Before the Wave

Every revolution creates winners and losers. The winners aren't the ones who scramble once the wave hits — they're the ones already paddling before it breaks.

The loom changed weaving. The personal computer changed office work. The internet changed commerce. Each time, a window existed where building early meant building a position that lasted. The operators who recognized the window and acted on it rode the wave. The ones who waited got buried under it.

AI is that wave. The flood of new competition is the tide that precedes it. And your Six-Layer Moat — reviews, brand, visibility, contracts, community, efficiency — is what keeps your customers choosing you regardless of how crowded the market gets.

You've read the whole book. You have the P&L framework. The pricing discipline. The customer retention system. The marketing infrastructure. The hiring foundation. The technology stack. The AI tools. The visibility strategy. Everything that makes a service business hard to compete with is in these pages.

The only variable left is whether you build it.

Start now. The wave doesn't wait.

Go Deeper

For more on this and everything else in the book, the Haulers' Edge Newsletter goes deeper every week. Scan the code or subscribe free at HaulingHubb.com.

CH. 37 WHAT WINNING ACTUALLY LOOKS LIKE

"The purpose of a business is to create a customer who creates customers." — Shiv Singh

Nobody tells you what winning looks like in a service business.

The business books describe it in abstractions — financial freedom, passive income, time independence. The Instagram highlight reels show beach houses and laptops by the pool. The podcast guests talk about "exiting" or "scaling to eight figures."

None of that tells you what Tuesday feels like when the business is actually working.

This chapter is about that Tuesday. About what you've built when the principles in this book are operational and compounding. Not the fantasy version — the real one. The one that's actually achievable for a service business owner who executes consistently over several years.

Because here's what I've learned running businesses in the trenches: winning doesn't arrive as a single moment. It arrives as a slow accumulation of small things working the way they're supposed to. And one day you look up and realize the business feels different. That's what this chapter is about.

What Winning Looks Like Financially

Winning financially doesn't mean you never think about money. It means money has stopped being the source of constant anxiety it used to be.

You know your numbers. You can tell someone your gross margin, your net margin, your cost per job, and your break-even point without looking anything up. You review your P&L monthly and it tells you a story you can act on. Your books are clean. Tax season is a payment, not a crisis, because you've been setting aside 25% all year.

Your cash flow is predictable enough that you're not scrambling at the end of every month. You have a reserve — not a massive one, maybe two to three months of operating expenses — that takes the existential edge off a slow week or an unexpected repair. The reserve didn't appear overnight. You built it intentionally, a percentage of revenue at a time, until one day it was there and the business felt structurally different because of it.

Profit is real. Not just revenue that looks impressive but actual money left after every legitimate cost has been paid — labor, disposal, fuel, overhead, your own salary taken properly. When you do a job, you know roughly what you'll keep from it. You've run the job costing math from Chapter 3 enough times that the numbers are intuitive. You're not guessing. You're tracking.

Winning financially also means you've made good decisions about what to do with profit. Some of it went back into the business — equipment that improved efficiency, technology that captured leads you used to lose, people who made the operation more capable. Some of it went into your personal financial picture — loans paid down, savings built, a retirement contribution that finally started. The business generates enough that you're not stuck choosing between investing in growth and taking care of yourself.

You don't think about money the way you used to. It's still important — it always will be — but it's become a tool rather than a source of dread. That shift is real and it's significant. Most people never get there. You did.

What Winning Looks Like Operationally

Winning operationally means the business runs with or without you in every job.

Your crew knows what to do. Not because you're there telling them — because you built the SOPs, ran the training, and hired people with enough character that they do the right thing when you're not watching. The job arrival process, the completion walkthrough, the customer interaction — all of it happens consistently because it's documented and practiced, not because you're on-site supervising.

The phone gets answered. Not always by you. Sometimes by your admin, sometimes by the AI voice agent, sometimes by a crew member who knows the intake process. But it gets answered, and the lead gets captured, and the follow-up happens — because you built the system that makes that automatic instead of dependent on your personal availability.

Your schedule has reliable structure. A meaningful percentage of your weekly revenue is from recurring commercial clients who call because you're part of their workflow, not because they found you on Google that day. The feast-and-famine cycle that dominated the early years has stabilized. You know roughly what next month looks like before it starts.

You can take a week off. Actually off — not checking in twice a day, not fielding calls from the crew, not coming back to a disaster. The business runs for a week without you and when you return the reports show a normal week. That capability — the business functioning independently of your daily presence — is one of the most concrete expressions of operational winning you can achieve.

It didn't happen because you hired perfectly or because you got lucky. It happened because you documented your processes, built your team deliberately, and invested in the systems that run without you. The week you take off is the proof of that work.

What Winning Looks Like in the Market

Winning in your market means your name means something before anyone calls you.

You have a review profile that does your selling before the phone rings. When a homeowner in your service area searches for what you do, your 200-plus reviews at a 4.8 average make you the obvious call. The customer arriving at your website or your GBP has already decided they're probably going to hire you — they just need to confirm you're available. That's a fundamentally different conversion dynamic than competing on an even playing field for every job.

Your trucks are recognized. Contractors know your name without looking you up. Property managers have you in their phones and call without getting quotes from anyone else. You've become part of the infrastructure of the communities you serve — not just a vendor, but a known, trusted presence. That recognition took years to build. Now it compounds.

You win jobs without bidding for them. Not every job — but enough that your close rate on inbound leads is high because your reputation is doing pre-selling work. Customers who call you have usually already decided you're the right choice. The call is to confirm availability and get the price, not to shop you against two competitors.

When new competitors enter your market — and they will — their first challenge is your review profile. Their second challenge is that your commercial clients already have vendors they trust. Their third challenge is that your name is visible everywhere in the market while they're building from zero.

None of those things are insurmountable, but they're real friction that buys you time and protection.

What Winning Looks Like in Your Life

Here's the version nobody talks about enough.

Winning in your life means the business serves your life rather than consuming it.

You're present for your family in a way that wasn't possible in the early years. Not perfectly — the business still demands things, there are still hard weeks, there are still problems that need your attention. But the constant state of overwhelm, the never-leaving-work feeling, the anxiety that followed you home every night — that's gone or significantly diminished. The systems you built created enough space that you can close the laptop at a reasonable hour without feeling like you're letting the business down.

You have perspective on the business that you couldn't access when you were in survival mode. You can see patterns in the numbers. You can think strategically about where the business is going rather than just managing the emergency in front of you. You have mental bandwidth for the decisions that actually matter — who to hire next, which market to expand into, whether a partnership makes sense — because the operational baseline runs without constant input from you.

You've built something real. Not a job you own. A business. Something with systems and a team and a brand and recurring revenue and a reputation in the market. Something that exists as an entity separate from you, capable of operating and growing with your guidance rather than your constant presence. If you chose to sell it, it would be worth selling. If you chose to step back, it would keep running. If you chose to hand it to someone else to operate, they could.

That's the version of winning that this book was designed to help you build. Not the abstract financial freedom concept — the actual operational reality of a business that works, that you own, that serves your life.

What You Do With Profit Once You Have It

The financial discipline that makes winning sustainable — cash reserves, smart reinvestment, debt reduction, paying yourself correctly, and building toward a specific goal — is in Chapter 29. If you skipped that chapter, go back. Those rules are what keep the win from unraveling.

The Version of This Business Worth Building

I started Grizzly Junk Pros in 2014 with a truck and no clear picture of what I was building toward. I learned the hard way what this book tries to teach the easier way — the financial principles, the operational systems, the marketing infrastructure, the people decisions, the technology stack. Every chapter in this book represents something I either got right eventually or got wrong first.

The business I have today is different from the one I had in year two. Not just bigger — structurally different. It has recurring commercial revenue that I don't have to re-earn every month. It has systems that run when I'm not watching. It has a team that knows what good looks like and delivers it consistently. It has a review profile that does work before anyone calls. It has the technology infrastructure to capture and convert leads at every hour. It has financial clarity that lets me make decisions from data rather than gut feel.

Is it perfect? No. There are still problems, still challenges, still things I want to improve. But it's a business, not a job I own. And the distinction between those two things is the whole game.

That's what this book was always about. Not the tactics individually — the tactics are in service of that single transformation. From the operator who is consumed by the business to the operator who has built a business that serves them.

The financial literacy from Part 1. The marketing and lead flow systems from Part 3. The operational infrastructure from Part 4. The recurring revenue and protection from Part 5. The AI advantage from Part 6.

All of it builds toward Tuesday. The Tuesday where the schedule is full, the team is running jobs well, the follow-ups are firing automatically, the reviews are coming in without you asking, the commercial clients are calling without you chasing, and you're spending your time on the decisions that actually move the business forward rather than on the operational fires that used to own your day.

That Tuesday is real. It's built, not found. And everything you need to build it is in this book.

Now Go Build It

The operators who read business books and don't change anything outnumber the ones who read and act by a significant margin. Most of the value in this book will never be captured because the people who read it will nod along, agree with the principles, and return to operating exactly as they did before.

Don't be that person who reads and doesn't act.

Pick one chapter that addresses the most painful thing in your business right now. Not the most interesting chapter — the most necessary one. The cash flow problem from Chapter 2. The pricing gap from Chapter 5. The missed calls from Chapter

16. The follow-up system from Chapter 17. The review deficit from Chapter 10. Whatever is costing you the most money or the most peace of mind right now — start there.

Everything in this book requires focused execution to work. The operators who implement half of it while chasing the next business idea, splitting attention across multiple ventures, or mentally planning the next move before this one runs itself — they get partial results at best. The compounding that makes these systems powerful only activates when you go all in on one thing long enough for it to actually compound. The most dangerous distraction isn't a bad decision. It's a good opportunity at the wrong time. Build this business first. Build it completely. Then expand.

Before you close this book — one mindset shift that's changed how I think about goals. When I set a goal that's actually achievable from where I currently stand, I don't change anything fundamental. I just do more of what I'm already doing. The goal is close enough that my existing habits can get me there. Which means I stay the same operator, just slightly bigger.

The goals that actually pushed me to build differently were the ones that couldn't be reached with what I had. When the number's too big for your current operation to touch, you can't get there by working harder — you have to build differently. That question — *what would actually have to change?* — cuts through the noise faster than any planning session. It shows you what matters for the business you're trying to build, not just the slightly larger version of the one you already have.

Write down one goal that genuinely feels out of reach at your current scale. Then ask — not how do I get there eventually — but what would have to be completely different for this to happen fast? That answer is your actual strategy.

Implement that one thing fully before you move to the next.

Not dabble with it. Build it, run it, measure it, refine it until it's operating the way it should. Then go to the next chapter that addresses the next most important problem.

Six months of that approach produces a business that looks meaningfully different. Twelve months produces one that feels structurally different. Three years produces the business described in this chapter — the one where the systems run, the revenue is recurring, the market knows your name, and the business serves your life rather than consuming it.

The whole game is in the execution. The knowledge in this book is available to every operator in your market. The ones who win are the ones who build it.

That's you. Go.

Go Deeper

The progress, the setbacks, and the honest in-progress version of everything in this book — that's what the Haulers' Edge Newsletter covers every week. If you want to keep building alongside someone still in it, that's where to find me. Scan the code or subscribe free at HaulingHubb.com.

EPILOGUE

The real trap isn't chains you can see. It's the conditioning you never question. From the time we're young, we're taught to trade hours for dollars, to believe debt is normal, to think the safe path is the only path. That conditioning becomes the invisible cage most people live their whole lives inside.

In business, that means rejecting the script that says you have to underprice to win jobs, hustle endlessly without profit, or stay small because growth feels risky.

The truth is, you don't have to accept the default. You can reframe, rebuild, and create your own rules.

Freedom comes from questioning everything you were told and then learning, testing, and carving your own path — without being beholden to industry norms.

That's how you buy back your time, your energy, your life.

That's how you go beyond breaking even.

—*Justin Hubbard*

ABOUT THE AUTHOR

Justin Hubbard started Grizzly Junk Pros in 2014 with a truck and a willingness to outwork everyone around him. What began as a local junk removal operation in Connecticut has grown into a company serving over 16,000 customers, running six trucks and more than 100 twenty-cubic-yard roll-off dumpsters, and completing over 6,000 jobs a year — without Justin's day-to-day involvement. Building a business that runs without its founder is a milestone most operators never reach. It became his obsession.

Along the way he invented the Grizzly Bag, a proprietary dumpster bag pickup service, acquired two hauling companies, launched and sold a gutter cleaning business, and founded New Life Warehouse to keep quality items out of landfills.

Justin leads Adimize, a digital marketing agency built specifically for service businesses, offering paid ads, AI Search Visibility SEO, websites, CRM automation, AI voice and text agents, and reputation management. He also created the Haulers' Edge AI — a custom GPT trained on over twelve years of hauling and demolition expertise that gives operators instant photo-to-quote estimates, profit calculations, SOP generation, and on-demand business coaching.

Every week he publishes The Haulers' Edge Newsletter, where he shares what he's testing, what's working, and what isn't — in real time. He hosts the Smokin' Profits Podcast, where he breaks down what's actually working in service businesses right now.

He lives in Tampa, Florida with his wife and their soon-to-be four children. His core belief — that a business should serve your life, not consume it — is what this book is built on.

Connect with Justin:

GLOSSARY

Terms and phrases used throughout this book, defined in plain language for quick reference.

80/20 Rule (Pareto Principle) The observation that in most businesses, roughly 80% of results come from 20% of inputs — 80% of revenue from 20% of customers, 80% of profit from 20% of service lines. The exact ratio varies, but the pattern is consistent: results are never evenly distributed. The practical application is identifying your highest-performing 20% and directing disproportionate energy toward it. *See Chapter 12.*

Accounts Receivable Money owed to your business for work already completed but not yet collected. An invoice you sent that hasn't been paid is accounts receivable. High or aging accounts receivable — invoices sitting unpaid for 30, 60, or 90+ days — is a common source of cash flow problems even in otherwise profitable businesses. *See Chapter 2.*

Accrual Accounting An accounting method that records revenue when it is earned and expenses when they are incurred, regardless of when money actually moves. A job completed in December is recorded as December revenue under accrual even if payment arrives in January. Contrast with cash basis accounting. Most small service businesses operate on cash basis. *See Chapter 1.*

Agentic AI AI that operates proactively toward a goal rather than waiting for a question. Unlike conversational AI — which responds when prompted — an AI agent receives a goal, determines the steps to accomplish it, executes those steps in sequence, and delivers a result. A research agent that builds a lead list overnight while you're not working is an example of agentic AI. *See Chapter 33.*

AI Search Visibility SEO An approach to local search optimization that goes beyond traditional website ranking to ensure your business is represented in AI-generated answers, voice search results, zero-click search results, and Google Business Profile displays. The goal is to be the business an AI recommends when a customer asks for a service recommendation in your market. *See Chapter 34.*

Asset Utilization Rate The percentage of your revenue-generating assets actively producing revenue versus sitting idle. For a dumpster operation: dumpsters on job sites divided by total dumpsters in your fleet. For a truck-based operation: billable truck days divided by available truck days. High, sustained utilization signals genuine capacity pressure and justifies adding assets. A seasonal spike followed by a slow-period drop does not. *See Chapter 19, Appendix A.*

Average Days to Collect The average number of days between completing a job and receiving payment, across all invoiced jobs in a given month. For residential work, this should be near zero if you collect on-site. For commercial accounts with net-30 or net-60 terms, it tracks how closely customers are paying to terms. A rising number is an early warning sign of cash flow problems before they become a crisis. *See Chapter 12.*

Average Job Value Total revenue for a given period divided by the number of jobs completed in that period. Tracks whether your pricing is holding steady or quietly drifting downward over time. A declining average job value while job count holds steady means your prices are eroding. *See Chapter 12.*

Balance Sheet A financial statement showing what your business owns (assets), what it owes (liabilities), and the difference between the two (equity). The P&L tells you whether you were profitable over a period. The balance sheet tells you what the business is actually worth at a point in time. Lenders review it when evaluating loan applications. Buyers review it when evaluating your business for acquisition. *See Chapter 1.*

Break-Even Point The amount of revenue your business must generate each month before any profit exists. Calculated as total monthly fixed overhead divided by gross profit margin. Every dollar of revenue above break-even contributes to profit. Every dollar below it is a loss. *See Chapter 1, Appendix C.*

Cash Basis Accounting An accounting method that records revenue when money is actually received and expenses when they are actually paid. The simplest and most common method for small service businesses. Contrast with accrual accounting. *See Chapter 1.*

Cash Flow The movement of money in and out of your business in real time. Distinct from profit — a business can be profitable on paper while simultaneously running out of cash if invoices aren't being collected, overhead is high, and payments are due before income arrives. Cash flow is what actually pays your bills. *See Chapter 2.*

Certificate of Insurance (COI) A document generated by your insurance broker that confirms your business coverage is in force. Commercial clients, property managers, and contractors typically require a current COI before allowing you on their job sites or adding you to their vendor lists. Subcontractors you hire should also provide their own COIs before performing any work. *See Chapter 28.*

Close Rate The percentage of estimates or leads that convert to booked jobs. Close rate overall (all leads) divided by source (Google Ads, referral, etc.) is more useful because different lead sources convert at very different rates. *See Chapters 16, 17, 12.*

Compounding The process by which consistent, incremental improvements generate returns that build on themselves over time. A customer retained generates a referral who generates a review who attracts a new customer — each cycle building on the last. The same principle applies to reviews, brand recognition, and operational efficiency. *See Chapters 8, 9, 11.*

Conversational AI AI tools that handle a wide range of language and reasoning tasks through natural dialogue — answering questions, drafting content, analyzing information, and generating responses. Examples include Claude, ChatGPT, and Gemini. Distinct from narrow AI (which does one task, like bid optimization) and agentic AI (which takes autonomous sequences of actions toward a goal). *See Chapter 30.*

Cost Per Acquired Customer Total marketing spend on a channel divided by the number of customers actually acquired from that channel. Different from cost per lead — a cheap lead that never books is not a cheap customer acquisition. The distinction matters for comparing the true ROI of different marketing channels. *See Chapter 12.*

CRM (Customer Relationship Management) Software that tracks every customer interaction, manages your sales pipeline, runs automated follow-up sequences, and stores the complete history of every contact. The operating system of your customer-facing business. A properly configured CRM ensures no lead falls through the cracks and every customer relationship is systematically maintained. *See Chapters 11, 17, 21.*

Customer Lifetime Value (CLV) The total revenue a customer generates over their entire relationship with your business — not just the first job. A customer who books once and never returns has a lifetime value equal to that single job. A customer who books annually for ten years plus refers three neighbors has a much higher lifetime value. Understanding CLV by lead source changes how you think about marketing spend. *See Chapter 12.*

Depreciation The gradual reduction in value of a physical asset over its useful life. A $60,000 truck that will last 150,000 miles has a depreciation cost of $0.40 per mile — a real cost that reduces profit whether or not it shows up on an invoice. Failing to account for depreciation in job costing leads to underpriced work and no money for repairs or replacement when the asset eventually fails. *See Chapters 3, 6.*

Direct Costs (Cost of Goods Sold / Cost of Services) The expenses directly tied to completing a specific job — labor, fuel, disposal fees, materials, equipment depreciation. These costs exist because the job exists; they go up when work increases and down when work slows. Distinct from overhead, which exists regardless of job volume. *See Chapter 1.*

EBITDA (Earnings Before Interest, Taxes, Depreciation, and Amortization) A measure of a business's core operating profitability, stripping out financing, tax, and accounting variables to allow comparison between businesses. The primary valuation metric buyers use when evaluating a service business for acquisition. Not a metric for daily or monthly tracking, but critical to understand if you're within five years of a potential exit. *See Chapter 29, Appendix A.*

Equity The net worth of your business — what's left when you subtract total liabilities from total assets on the balance sheet. Growing equity means the business is building real value over time. Negative or shrinking equity means debt is outpacing assets. *See Chapter 1.*

Expense-to-Revenue Ratio Total expenses in a period divided by total revenue in the same period. A quick pulse check on whether your cost structure is staying in proportion to your revenue. Any significant unexplained climb in this ratio is a signal to investigate. *See Chapter 12.*

Fixed Overhead Business expenses that continue at a fixed amount regardless of how many jobs you run — vehicle payments, insurance, software subscriptions, rent, salaries. These costs exist whether you complete one job this month or a hundred. They set your break-even floor and do not decrease automatically during slow periods. *See Chapter 1.*

Follow-Up Sequence A structured, multi-touch series of contacts — texts, emails, calls — sent to a lead or estimate that has not yet converted. Industry data consistently shows that most sales require multiple follow-up attempts, yet most service businesses make only one. An automated follow-up sequence runs the cadence without manual intervention. *See Chapter 17.*

Google Business Profile (GBP) Your free business listing on Google that controls how your business appears in Google Maps, local search results, and AI-generated local recommendations. One of the highest-return digital assets available to a local service business at no cost. Review count, recency, completeness, and posting frequency all affect your ranking. *See Chapters 9, 13, 34.*

Gross Margin (Gross Profit Margin) The percentage of revenue remaining after direct job costs — before overhead, taxes, or owner pay. Calculated as gross profit divided by revenue. A 50% gross margin means 50 cents of every dollar earned from jobs is available to cover overhead and generate profit. Most service businesses should target 45–60% gross margins. *See Chapters 1, 3, 5.*

Gross Profit Revenue minus direct job costs. What's left from a job after paying for the labor, fuel, disposal, and materials required to complete it — before overhead is accounted for. *See Chapter 1.*

Inbound Lead A prospective customer who contacts your business on their own initiative — calling, filling out a form, or messaging — as opposed to outbound outreach where you initiate contact. Most residential service business revenue comes from inbound leads generated through Google search, Google Business Profile, referrals, and yard signs. *See Chapters 14, 16.*

Job Costing The practice of calculating the true total cost of a specific job before pricing it — accounting for direct labor with burden, owner time, fuel, disposal fees, materials, equipment depreciation, and overhead allocation. The foundation of accurate pricing. *See Chapter 3, Appendix B.*

Labor Burden The total cost of an employee beyond their base wage — including payroll taxes, workers' compensation insurance, and any benefits. Typically adds 15–25% on top of base wages. A $20/hour employee costs approximately $23–25/hour in real cost once burden is included. Failing to account for labor burden leads to systematic underpricing. *See Chapter 3.*

LLC (Limited Liability Company) A business entity structure that creates a legal separation between your personal assets and your business liabilities. A single-member LLC is taxed as a sole proprietorship by default but provides the personal liability protection a sole proprietorship does not. The minimum structure recommended for any service business operating trucks, working in customers' homes, or employing people. *See Chapter 6.*

Local Services Ads (LSAs) A Google advertising format that appears above standard Google Ads in search results, featuring a Google Verified badge, your review rating, and your business name. Pay-per-lead rather than pay-per-click. Requires background check and license verification. For qualifying service businesses, LSAs often deliver lower cost per lead than standard search campaigns with higher placement. *See Chapter 14.*

Merchant Cash Advance (MCA) A form of short-term business financing where a lender provides a lump sum in exchange for a percentage of future revenue. Often marketed to small businesses as fast and accessible. Effective annual interest rates frequently exceed 40–100% when the full cost is calculated. Among the most expensive forms of small business financing available. *See Chapter 1.*

Moat A competitive advantage that is difficult and slow for competitors to replicate — used metaphorically, from the water barrier around a castle that slows attackers. In a service business, a moat consists of accumulated review volume, brand recognition, established commercial relationships, community integration, and operational efficiency. Each layer takes time to build. Combined, they create a position new entrants cannot quickly duplicate. *See Chapter 36.*

NAP Consistency The uniformity of your business Name, Address, and Phone number across every online platform where your business is listed. Inconsistencies in NAP — different phone formats, old addresses, slight name variations — create ambiguity that reduces local search rankings and AI search visibility. Your NAP must be identical everywhere, not just similar. *See Chapter 34.*

Net Profit (Net Income) What remains after all expenses — direct costs, overhead, debt service, taxes, and owner pay — are subtracted from revenue. The truest measure of business profitability. A business can have strong gross margins and still have weak or negative net profit if overhead is high or debt service is significant. *See Chapter 1.*

New Customer Velocity The count of first-time customers who booked in a given week or month — people who have never used your business before. Tracking this separately from total job count reveals whether growth is coming from new customers or simply from existing customers returning. A flat or declining velocity while total revenue holds steady signals a weakening acquisition pipeline. *See Chapter 12.*

Operating Margin Net profit divided by total revenue for a period. The single most useful summary number for overall business profitability — it accounts for everything including overhead, debt, and owner pay. Track it monthly and year over year. *See Chapter 12.*

Overhead Allocation The portion of your fixed monthly overhead assigned to each individual job, calculated by dividing total monthly overhead by average monthly job count. If your monthly overhead is $6,000 and you run 80 jobs, each job must contribute $75 to overhead before any profit exists. Most operators who omit overhead allocation from job costing consistently underprice their work. *See Chapter 3.*

Owner's Draw The method by which sole proprietors and single-member LLC owners pay themselves — by transferring money from the business account to their personal account. Unlike an S-Corp salary, an owner's draw does not show up as a payroll expense on the P&L, which means profit appears higher than what the owner is actually keeping after personal income. *See Chapter 1.*

P&L (Profit and Loss Statement / Income Statement) A financial statement showing revenue, direct costs, gross profit, overhead expenses, and net profit over a specific period — typically a month, quarter, or year. The primary document for understanding whether a business is profitable. Does not show cash position or asset health. *See Chapter 1.*

Pass-Through Entity A business structure where business income flows through to the owner's personal tax return rather than being taxed at the corporate level. LLCs and S-Corps are both pass-through entities. The advantage is avoiding double taxation (once at the corporate level and again when distributed to owners) that applies to C-Corps. *See Chapter 6.*

Payroll Burden *See Labor Burden.*

PITA Premium (Pain in the Ass Premium) A pricing surcharge applied to jobs that carry higher-than-average difficulty, risk, or hassle — tight access, unusual materials, difficult customers, complex logistics. The principle: jobs that make your operation work harder or carry more uncertainty should cost more. Also called a walk-away price when set high enough that declining the job feels acceptable if the customer says no. *See Chapter 4.*

Pipeline The organized tracking of all active leads and prospects through the stages of your sales process — from initial inquiry through estimate, follow-up, and booking. A pipeline managed in a CRM ensures every lead is tracked and no opportunity falls through the cracks due to volume or distraction. *See Chapter 21.*

Quality Score Google's rating of the relevance and quality of your paid search ads and landing pages, scored 1–10. Higher Quality Scores result in better ad placement at lower cost per click. Influenced by expected click-through rate, ad relevance to the search query, and landing page experience. *See Chapter 14.*

Recurring Revenue Revenue generated from customers or clients who engage your services repeatedly on an ongoing basis, without requiring a new sales process each time. Contractors who call monthly for debris removal, property managers who rely on you for every unit turnover, and standing service agreements all represent recurring revenue. More predictable, more defensible, and valued more highly by business buyers than transactional revenue. *See Chapter 27.*

Repeat and Referral Rate The percentage of jobs in a given month that came from past customers or their referrals, as opposed to new cold leads. A growing repeat and referral rate means your retention and relationship systems are working. A flat or declining one — even while total revenue holds — means you're replacing customers you should be keeping. *See Chapters 11, 12.*

Revenue Per Crew Hour Total revenue for a period divided by total billable field hours worked by crews in that period. The single most useful operational efficiency metric for a service business. Establishes a baseline when the operation is running well and flags problems — overburdened jobs, thin-margin work, inefficient routing — when the number drops. *See Chapters 12, 19.*

Route Optimization The practice of sequencing daily jobs geographically to minimize total drive time, fuel cost, and unnecessary movement between stops. Reduces per-job operating costs, allows more jobs per day with the same fleet, and decreases crew fatigue. Can be done manually with basic mapping tools or automated with dedicated routing software. *See Chapter 19.*

S-Corp Election A tax classification available to LLCs and corporations that allows business income to be split between a reasonable owner salary (subject to payroll taxes) and owner distributions (not subject to self-employment tax). Typically produces meaningful tax savings when business net profit exceeds $50,000–$60,000 annually. Requires running actual payroll and paying a defensible salary. *See Chapter 6.*

Schema Markup Code added to a website that explicitly tells search engines and AI systems what a business does, where it operates, what services it offers, and what customers say about it. Helps search algorithms and AI recommendation systems accurately represent your business without having to infer details from unstructured text. *See Chapter 34.*

Secret Shopping The practice of contacting competitors as a prospective customer to experience their sales process firsthand — how they answer the phone, how they quote, how they follow up. Provides competitive intelligence that is both honest and legal. Recommended quarterly as part of a systematic competitive review. *See Chapter 7.*

Section 179 An IRS tax provision that allows businesses to deduct the full cost of qualifying equipment in the year it is purchased rather than depreciating it over time. Useful for timing large equipment purchases to reduce taxable income in high-revenue years. Limits and eligible property types adjust annually — verify current rules with your CPA. *See Chapter 6.*

Self-Employment Tax The tax covering both the employer and employee share of Social Security and Medicare for self-employed individuals and business owners. Currently 15.3% on net profit. As a W2 employee, your employer pays half. As a sole proprietor or standard LLC member, you pay all of it. One of the primary financial reasons for the S-Corp election at higher income levels. *See Chapter 6.*

Six-Layer Moat A competitive defense framework for service businesses consisting of six stackable layers: review dominance, brand recognition, AI Search Visibility SEO, commercial contracts, community integration, and operational efficiency. Each layer is difficult for a new market entrant to replicate quickly; stacking all six creates a compounding position that protects market share through saturation cycles. *See Chapter 36.*

SOP (Standard Operating Procedure) A documented, step-by-step description of how a specific task or process should be performed in your business. SOPs eliminate quality variation that depends on who is working, capture institutional knowledge so it survives staff turnover, reduce training time for new hires, and create a foundation for accountability. *See Chapter 20.*

Straight-Line Depreciation The simplest method of calculating depreciation, dividing an asset's cost evenly across its estimated useful life. A $60,000 truck over 150,000 miles of useful life depreciates at $0.40 per mile. Useful for incorporating depreciation into job costing. *See Chapter 3.*

Switching Cost The friction, time, risk, and inconvenience a client would face if they replaced your business with a competitor. A property manager who has trusted you for three years, knows your crew, and has your number saved faces real switching costs that make them less likely to change vendors for incremental price differences. Building switching costs through reliability, systems, and relationships is a core competitive advantage. *See Chapter 27.*

Three-Touch System A structured follow-up sequence for unconverted estimates consisting of three contacts over thirty days: a Day 1-2 confirmation touch, a Day 5-7 value touch, and a Day 30 door-open close. Designed to convert leads that went quiet after receiving an estimate without requiring manual tracking or daily attention from the operator. *See Chapter 17.*

True Job Cost The complete, accurate cost of performing a specific job — including direct labor with burden, owner time, fuel for the full route, disposal fees, materials and supplies, equipment depreciation, and an overhead allocation. The floor below which any job loses money. Calculating true job cost before pricing is the foundation of sustainable profitability. *See Chapter 3, Appendix B.*

Two-Path Review System A post-job follow-up system that routes customers through a single satisfaction question before directing them to a public review platform. Satisfied customers are sent to Google (or other review platforms). Unsatisfied customers are routed to a private internal feedback form — giving them a channel to be heard while allowing you to resolve the issue before it becomes a public review. *See Chapter 10.*

Variable Costs Costs that rise and fall with business volume — fuel, disposal fees when volume increases, subcontractor labor called in for busy periods. Unlike fixed overhead, variable costs do not exist unless jobs are running. Some businesses have significant variable costs that must be accounted for in break-even calculations beyond the basic fixed overhead formula. *See Chapter 1.*

Valuation Multiple The factor applied to annual profit (or EBITDA) to determine a business's sale price. A business with $200,000 in annual profit selling at a 3x multiple is worth $600,000. Service businesses where the owner is essential to daily operations typically sell at 1–2x. Businesses with strong systems, documented processes, recurring revenue, and owner-independent operations can sell at 3–5x or higher. *See Chapter 29.*

Winning Call Framework A five-stage structure for inbound service business calls: (1) Answer fast and set the tone, (2) Understand the job before quoting, (3) Quote with confidence and clarity, (4) Handle the objection without flinching, (5) Close and confirm. Designed to convert inbound calls consistently without relying on individual sales instinct. *See Chapter 16.*

Zero-Click Search A search result where the user gets the information they need directly on the results page — from an AI-generated summary, a featured snippet, a Google Business Profile, or a map pack listing — without clicking through to any website. A growing majority of local service searches end without a website visit. Being present in zero-click results requires a different optimization strategy than traditional website SEO. *See Chapter 34.*

APPENDIX A: THE BUSINESS METRICS SCORECARD

Your weekly and monthly operating dashboard. The 80/20 audit in Chapter 12 tells you where your business has been. This scorecard tells you where it's going right now.

The weekly review takes 30 minutes. The monthly review takes 60–90 minutes. Neither requires a financial background. Both require consistent data entry in your CRM and a standing appointment with yourself — same day, every week, every month.

The P&L tells you what happened. The scorecard tells you what's happening now, while you still have time to act on it.

Weekly Scorecard — 30 Minutes

Run this every week, same day.

Total Qualified Leads by Source Count of new leads received this week, tagged by source in your CRM.

How to calculate: Pull from your CRM pipeline. Count every new lead that entered this week. Break the count down by source — Google Ads, referral, Nextdoor, repeat customer, yard sign, etc.

Why it matters: This number is only meaningful if every lead is tagged at entry. Without source tagging, you're looking at volume with no context and every downstream metric loses its meaning. Build the tagging habit before you need the data.

This week's count: ___ By source: Google Ads ___ / Referral ___ / Repeat ___ / Other ___

Ad Spend by Source Total dollars spent this week on each paid channel.

How to calculate: Pull weekly spend from each ad platform — Google Ads, LSAs, Facebook, Nextdoor, any others. Record separately.

Why it matters: The input number that makes every other marketing metric meaningful. Close rate and cost per customer mean nothing if you don't know what you spent to generate the lead.

This week: Google Ads $___ / LSAs $___ / Other $___

Estimate Close Rate — Overall Jobs booked this week divided by estimates sent this week.

How to calculate: (Jobs booked ÷ Estimates sent) × 100

Why it matters: Your baseline conversion number. A declining close rate with stable lead volume means your phone process, follow-up, or pricing is losing ground somewhere.

This week: ___ booked ÷ ___ estimates = ___%

Close Rate by Source The same calculation run separately for each lead source.

How to calculate: For each source, divide jobs booked from that source by estimates sent to leads from that source.

Why it matters: Your Google Ads close rate and your referral close rate will be very different numbers — and that difference is where your budget decisions should come from, not total lead volume.

Google Ads: ___% / Referral: ___% / Repeat: ___% / Other: ___%

New Customer Velocity Count of first-time customers who booked this week — people who have never used your business before.

How to calculate: Filter your CRM for new bookings this week and count those tagged as first-time customers.

Why it matters: Track this separately from total jobs. If total revenue holds but this number is declining, your growth is coming entirely from repeat business and your new customer pipeline is quietly weakening. That shows up here before it shows up anywhere else.

This week: ___

Total Jobs by Service Category Raw count of jobs completed this week in each service line.

How to calculate: Pull completed jobs from your CRM or scheduling platform, broken out by service type.

Why it matters: Not a comparison — just the number. This tells you volume and capacity utilization per service, which no other metric shows you directly.

Service 1 (___): ___ / Service 2 (___): ___ / Service 3 (___): ___

Total Revenue by Service Category Raw weekly revenue per service line.

How to calculate: Sum revenue from completed and invoiced jobs this week by service category.

Why it matters: Tells you which services are carrying the week and whether your mix is shifting — information the year-over-year comparison alone can't provide.

Service 1 (___): $___ / Service 2 (___): $___ / Service 3 (___): $___

Average Job Value Total revenue this week divided by number of jobs completed this week.

How to calculate: Total weekly revenue ÷ Number of completed jobs

Why it matters: Tells you whether your pricing is holding steady or quietly drifting in the wrong direction. A declining average job value with stable job count means you're doing the same work for less money.

This week: $___ ÷ ___jobs = $___

Revenue Per Crew or Technician Hour Total revenue this week divided by total billable field hours worked by your crew this week.

How to calculate: Total weekly revenue ÷ Total billable crew hours in the field

Why it matters: The single best operational efficiency number available to a service business. A junk removal operation running $8,000 in weekly revenue across 160 crew hours produces $50 per hour. A cleaning company, landscaper, HVAC technician, or plumber runs the same calculation. Set your baseline when the business is running well. Any week it drops significantly, something changed — jobs are running long, you're booking thin-margin work, pricing slipped, or crew efficiency fell off. Any week it climbs, something is working and you want to know what so you can repeat it.

Baseline (set when running well): $___/hour This week: $___ ÷ ___ hours = $___/hour

---

Asset Utilization Rate The percentage of your revenue-generating assets actively working versus sitting idle.

How to calculate:

- Dumpster operation: Dumpsters on job sites ÷ Total dumpsters in fleet
- Truck-based operation: Billable truck days ÷ Available truck days

Why it matters: Tells you whether you're ready to grow. Sustained high utilization over multiple weeks signals genuine capacity pressure and justifies adding assets. A seasonal spike followed by a slow-period drop is normal and does not signal the need for more capital. Don't let one busy month drive a decision that commits you to costs you'll carry all year.

This week: ___ active ÷ ___ total = ___%

No-Show and Cancellation Rate Cancellations and no-shows this week divided by total booked jobs this week.

How to calculate: (Cancellations + No-shows) ÷ Total booked jobs × 100

Why it matters: A rate above 10–12% is worth investigating—it's usually a booking process problem, a confirmation gap, or a customer segment issue. This metric also tells you whether your automated reminder system is working. Without a baseline before you implement reminders, you can't measure the improvement after.

This week: (___ + ___) ÷ ___ = ___%

Visibility Metrics Leading indicators that predict your lead volume two to four weeks from now.

How to calculate: Pull from Google Business Profile Insights and your website analytics platform.

Metrics to track:

- GBP views this week: ___
- Calls or messages from GBP: ___
- Website sessions: ___
- Website conversion rate (contacts ÷ sessions): ___%

Why it matters: A drop in GBP views or website sessions today means fewer inbound leads next week. You only notice the lead drop and have no idea why. Watching these upstream tells you why before it costs you jobs.

Review Velocity Number of new Google reviews received this week.

How to calculate: Check your Google Business Profile and count new reviews since last week.

Why it matters: Two to three new reviews per week is the target for a business actively building its local presence. Below that, your review profile is aging relative to competitors who are generating reviews consistently.

This week: ___ Running 4-week average: ___

Cash Position vs. Last Week Current account balance minus last week's balance.

How to calculate: Check your business bank account balance.

Subtract last week's recorded balance.

Why it matters: Directional. Are you moving the right way? A declining cash position despite healthy revenue usually points to a collection problem, a payables timing issue, or an overlooked expense.

Last week: $___ This week: $___ Change: $___

Expense-to-Revenue Ratio Total expenses this week divided by total revenue this week.

How to calculate: Total outflows this week (all expenses paid) ÷ Total revenue collected this week

Why it matters: Flag any week where this climbs above your baseline without an obvious explanation. A rising ratio without a corresponding explanation means something is leaking.

Baseline ratio: ___% This week: $___expenses ÷ $___revenue = ___%

Monthly Review — 60 to 90 Minutes

Run this in the first week of every month, covering the prior month.

Operating Margin Net profit divided by total revenue for the month.

How to calculate: Net profit ÷ Total revenue × 100

Why it matters: The truest single profitability number your business produces — it covers everything, including direct costs, overhead, debt service, and your own pay. Track it month over month and year over year. This is the number that tells you whether the business is actually healthy, not just busy.

Last month: $___ net profit ÷ $___ revenue = ___% Prior month: ___% Same month last year: ___%

Repeat and Referral Rate Jobs this month sourced from past customers or their referrals divided by total jobs this month.

How to calculate: (Repeat customer jobs + Referral jobs) ÷ Total jobs × 100

Why it matters: If this number is growing, your retention system is compounding. If it's flat or declining, you have a problem even if total revenue looks fine — you're replacing customers you should be keeping.

Last month: ___ repeat/referral ÷ ___ total = ___% Trend (3-month average): ___%

New Customer Velocity Trend Compare this month's new customer count to the prior three months.

How to calculate: Count new first-time customers who booked this month. Compare to the monthly average of the prior three months.

Why it matters: A flat or declining trend while total revenue holds is a warning sign — you're becoming increasingly dependent on repeat business with a shrinking pool of new customers entering the funnel.

This month: ___ 3-month average: ___ Trend: Growing / Flat / Declining

---

Total Jobs by Service Category — Monthly Raw job count per service line this month versus last month and versus the same month last year.

How to calculate: Pull from your CRM or scheduling platform by service type.

Why it matters: Your volume picture by service — which lines are growing, flat, or shrinking.

Service	This Month	Last Month	Same Month Last Year
___	___	___	___
___	___	___	___
___	___	___	___

Total Revenue by Service Category — Monthly Dollar-weighted version of the job count comparison above.

How to calculate: Sum revenue by service category for the month and compare to prior month and prior year.

Why it matters: Tells you whether growth in one service is masking a decline somewhere else. These two lines together — job count and revenue by service — tell you not just whether a service is growing but whether it's growing profitably. Volume up, revenue flat means average job value is declining. Both up means the service is genuinely healthy.

Service	This Month	Last Month	Same Month Last Year
___	$___	$___	$___

___	$___	$___	$___
___	$___	$___	$___

Year-Over-Year Job Count and Revenue — Total Business The macro view.

How to calculate: Compare total jobs completed and total revenue this month to the same month last year.

Why it matters: Is the business bigger than it was this time last year, and by how much? Month-over-month comparisons are noisy with seasonality. Year-over-year is the cleanest trend line.

Total jobs: This month ___ / Same month last year ___ / Change ___% Total revenue: This month $___ / Same month last year $___ / Change ___%

Average Days to Collect Average number of days between job completion and payment received, across all invoiced jobs this month.

How to calculate: For each invoiced job this month, calculate days from completion date to payment received date. Average across all invoiced jobs.

Why it matters: For residential work this should be near zero if you're collecting on-site. For commercial accounts it reflects your payment terms and collections discipline. A number climbing above 15–20 days on commercial accounts is an early warning sign that shows up here before it becomes a cash flow crisis.

This month: ___ days Prior month: ___ days

Cost Per Acquired Customer by Source Total spend on each marketing channel divided by customers actually acquired from that channel.

How to calculate: Channel spend ÷ Customers booked from that channel. Not cost per lead — cost per customer.

Why it matters: Close rates vary significantly by source, so a cheap lead from one channel may produce a more expensive customer than a costlier lead from another. This is the number that determines true marketing channel ROI.

Google Ads: $___ spend ÷ ___ customers = $___ per customer LSAs: $___ spend ÷ ___ customers = $___ per customer Other paid: $___ spend ÷ ___ customers = $___ per customer

---

Customer Lifetime Value by Source Average revenue per customer by source, multiplied by average rebook frequency for customers from that source.

How to calculate: Filter closed customers by source in your CRM. For each source group, calculate average total revenue per customer across all their jobs. Pull quarterly — it requires at least 90 days of CRM history to be meaningful.

Why it matters: This is where most operators' marketing assumptions fall apart. A Google Ads lead might close at a higher rate and average more per job — but if referral customers rebook three times more often, the lifetime value of a referral customer may be double despite the lower single-job value. This changes everything about where your budget should go.

Google Ads CLV (quarterly): $___ Referral CLV (quarterly): $___ Other source CLV (quarterly): $___

A Note on EBITDA Earnings Before Interest, Taxes, Depreciation, and Amortization is not a weekly or monthly tracking metric — it is too complex for routine calculation and requires clean accrual accounting to mean anything. But if you are within five years of a potential exit, your CPA should be calculating it annually.

EBITDA is the number a buyer uses to value your business. Ask your CPA for it once a year alongside your annual tax preparation. Know the number and track whether it is moving in the right direction year over year. Chapter 29 covers the exit picture in full.

Last year's EBITDA: $___ Prior year: $___ Trend: ___

How to Set This Up

Step 1: Source tagging. Before any of these metrics produce useful data, every lead entering your CRM must be tagged with its source at the moment of entry. Build this habit now, before you have enough data to analyze, so the numbers are there when you need them.

Step 2: Put it on the calendar. The weekly review happens on the same day every week — Friday afternoon or Monday morning work well for most service businesses. The monthly review happens in the first week of every month. Without a recurring appointment, it doesn't happen.

Step 3: Set your baselines. Run the first full week and month without judgment — just to establish your starting numbers. Baselines for revenue per crew hour, asset utilization, and close rate should be set during a week or month when operations are running at a solid normal pace. Neither your best week nor your worst.

Step 4: Track changes, not just numbers. A close rate of 45% means nothing without knowing whether it was 50% last month or 38%. The scorecard produces value through trends, not snapshots.

The operators who know their numbers make better decisions faster. Everything else in this book compounds more effectively when you can measure whether it's working.

APPENDIX B: JOB COST WORKSHEET

Use this worksheet on every job before you quote. The goal is simple: know your floor before you name a price. Build it into a spreadsheet once, use it every time.

The Formula

True Job Cost = Direct Labor + Owner Time + Fuel + Disposal Fees + Materials + Equipment Depreciation + Overhead Allocation

Gross Profit = Revenue − True Job Cost

Gross Margin % = Gross Profit ÷ Revenue

Your Price Floor = True Job Cost ÷ (1 − Target Margin)

Target margin for most service businesses: 40–50% on individual jobs. Higher on fast, simple work. Floor holds on complex jobs.

Line-by-Line Calculation

Direct Labor Number of crew members: ___ Hours on job (include drive time): ___ Hourly rate: $___ Labor subtotal: $___ Burden rate (payroll taxes + workers' comp, typically 15–20%): × ___ **Direct Labor Total: $___**

Note: Burden brings a $20/hr employee to roughly $23–24/hr in real cost. Use your actual burden rate, not the base wage.

Owner Time Hours spent (driving, working, coordinating,

quoting): ___ Your hourly value: $___ *If you don't know this number, divide your annual income goal by 2,000 working hours.* **Owner Time Total: $___**

Fuel Full route miles (yard → job → dump → yard or next job): ___ Miles per gallon loaded (most service trucks: 8–12 mpg): ___ Current fuel price per gallon: $___ Fuel cost = (miles ÷ mpg) × price per gallon **Fuel Total: $___**

Disposal Fees Estimated load weight or volume: ___ Transfer station rate (per ton or per load): $___ Material type premium (construction debris, appliances, tires, electronics): $___ **Disposal Total: $___**

Know your transfer station's rate card. Heavy or specialty materials cost significantly more. Build the actual estimate into every quote — not a rough guess.

Materials and Supplies Tarps, straps, bags, protective gear, cleaning supplies, etc.: $___ **Materials Total: $___**

Track this monthly. Most operators underestimate until they see the cumulative number.

Equipment Depreciation Vehicle cost: $___ Estimated useful life in miles: ___ Depreciation per mile = Vehicle cost ÷ Useful life miles = $___/mile Job miles: ___ Depreciation Total: $___

Example: A $60,000 truck over 150,000 miles = $0.40/mile. On a 40-mile round trip that's $16 in depreciation before maintenance, tires, or repairs.

Overhead Allocation Monthly fixed overhead (insurance, truck payments, software, marketing, phone, admin): $___ Average jobs per month: ___ Overhead per job = Monthly overhead ÷ Average jobs = $___ **Overhead Allocation: $___**

This is the most consistently skipped line. If you're not allocating overhead, your jobs aren't actually profitable — they're covering direct costs while overhead bleeds you out.

Summary

Line Item	**Amount**
Direct Labor	$___
Owner Time	$___
Fuel	$___
Disposal Fees	$___
Materials	$___
Equipment Depreciation	$___
Overhead Allocation	$___
True Job Cost	**$___**

Price Calculation

True Job Cost: $___

Target Gross Margin: ___%

Price Floor = True Job Cost ÷ (1 − Target Margin)

Price floor at 40% margin: True Job Cost ÷ 0.60 = $___

Price floor at 50% margin: True Job Cost ÷ 0.50 = $___

Market rate for this job type in your area: $___

Your quoted price: $___

Actual gross margin = (Quoted Price − True Job Cost) ÷ Quoted Price = ___%

Worked Example

Residential garage cleanout, quoted at $600.

Line Item	Calculation	Amount
Direct Labor	2 crew × 2.5 hrs × $20/hr × 1.20 burden	$120
Owner Time	1 hr × $75/hr	$75
Fuel	35-mile round trip, 10 mpg, $3.50/gal	$12
Disposal	One mid-weight load	$80
Materials	Tarps, bags	$10
Equipment Depreciation	35 miles × $0.40/ mile	$14
Overhead Allocation	$6,000/mo ÷ 80 jobs	$75
True Job Cost		**$386**
Gross Profit	$600 − $386	**$214**
Gross Margin	$214 ÷ $600	**36%**

This job lands at 36% — which clears a 35% minimum floor but falls short of the 40–50% target range covered in Chapter 3. If disposal runs heavier or the job takes an extra hour, it won't clear the floor. That's the point: running the numbers before you quote tells you exactly where the margin is tight and what variables can blow it.

When a Job Comes In Below Your Floor

Three choices — and only one should be routine:

1. **Reprice it** to clear your margin floor.
2. **Decline it.** Not every job is worth taking.
3. **Take it strategically** — knowing you're breaking even — because it leads to something more valuable (a recurring commercial relationship, a foothold in a new customer type). This option should be rare and deliberate, never accidental.

The operators who learn to say no to jobs below their floor are the ones who eventually stop being too busy to be profitable.

APPENDIX C: BREAK-EVEN CALCULATOR

Your break-even point is the amount of revenue you need to generate each month before a single dollar of profit exists. Calculate it once when your cost structure is set. Recalculate every time you add a fixed cost — a new truck, a new employee, a new lease.

The Basic Formula

Break-Even = Total Monthly Fixed Overhead ÷ Gross Profit Margin

---

Step 1 — Calculate Your Total Monthly Fixed Overhead

Fixed overhead is everything you pay whether you run one job or one hundred. It does not scale with volume.

Expense Category	Monthly Amount
Vehicle payments / leases	$___
Insurance (all policies, monthly equivalent)	$___
Equipment payments	$___
Software subscriptions (CRM, routing, accounting, etc.)	$___
Marketing baseline (website, SEO, any flat-rate services)	$___
Phone and communications	$___

Admin wages (if fixed, not per-job)	$___
Owner salary (what you pay yourself)	$___
Facility / storage rent	$___
Loan payments	$___
Other fixed costs	$___
Total Monthly Fixed Overhead	**$___**

Step 2 — Calculate Your Gross Profit Margin

Your gross profit margin is the percentage of each revenue dollar left after direct job costs — labor, fuel, disposal, materials, depreciation. Use your actual job data, not an estimate.

Pull your last 90 days of jobs:

Total Revenue (90 days): $___

Total Direct Job Costs (labor, fuel, disposal, materials, depreciation — but not fixed overhead): $___

Gross Profit = Revenue − Direct Costs = $___

Gross Profit Margin = Gross Profit ÷ Revenue = ___%

Most service businesses should be running 45–60% gross margins. Below 40% means your direct costs are too high relative to your pricing.

Step 3 — Calculate Basic Break-Even

Monthly Break-Even = Total Monthly Fixed Overhead ÷ Gross

Profit Margin

Example: $8,000 overhead ÷ 0.55 margin = **$14,545 monthly break-even**

Your calculation:

$___ overhead ÷ ___ margin = **$___ monthly break-even**

Every dollar of revenue above this number is profit. Every dollar below it is a loss.

Step 4 — Adjust for Variable Overhead (if applicable)

Some costs rise with volume but aren't tied to individual jobs — a part-time driver you call in on busy weeks, a seasonal worker, variable marketing spend. These reduce the margin available to cover fixed overhead.

If you have meaningful variable overhead costs:

Adjusted Break-Even = Fixed Overhead ÷ (Gross Profit Margin - Variable Cost Rate)

Variable cost rate = Variable overhead costs as a percentage of revenue

Example: $8,000 fixed overhead ÷ (0.55 gross margin - 0.10 variable cost rate) = $8,000 ÷ 0.45 = **$17,778**

That's $3,233 higher than the basic formula. For an operator running near break-even, that gap is the difference between thinking you're profitable and actually being profitable.

Your calculation (if applicable):

$___ fixed overhead ÷ (___ gross margin - ___ variable cost rate) = **$___ adjusted break-even**

Step 5 — Calculate Break-Even Per Job

Knowing your monthly break-even in revenue is useful. Knowing how many jobs it takes to get there is actionable.

Monthly Break-Even Revenue: $___

Average Job Value: $___

Jobs Needed to Break Even = Break-Even Revenue ÷ Average Job Value

$___ ÷ $___ = ___ jobs per month to break even

Everything above that number is margin. Below it is loss.

Seasonal Break-Even Check

Your break-even doesn't change with the seasons. Your overhead is the same in January as it is in July. Run this check to see how your slow season looks against your floor:

Slowest realistic monthly revenue (your slow season floor): $___

Monthly break-even: $___

Gap (or surplus): $___

If your slow season revenue falls below your break-even, you are running at a loss in those months and surviving on peak-season surplus. That's manageable if you plan for it — dangerous if you don't.

When to Recalculate

Recalculate your break-even immediately whenever:

- You add a truck, piece of equipment, or other fixed asset
- You hire a full-time or salaried employee
- You sign a new lease or long-term contract
- Your insurance or other fixed costs change significantly
- You raise or lower your prices
- Your service mix shifts and your average job value changes

Every new fixed cost raises your floor. Know by how much before you commit.

Summary

	Your Number
Total Monthly Fixed Overhead	$___
Gross Profit Margin	___%
Monthly Break-Even (basic)	$___
Monthly Break-Even (adjusted for variable overhead)	$___
Average Job Value	$___
Jobs Needed to Break Even	___
Slow Season Revenue	$___
Slow Season Gap / Surplus	$___

APPENDIX D: RECOMMENDED TOOLS AND RESOURCES

Every tool referenced in this book, organized by function. These are the platforms in active use at Grizzly Junk Pros or recommended based on track record in the service trades. Note: Service Hubb AI and Haulers' Edge AI are products the author built and operates. They're included because they're what runs at Grizzly Junk Pros — not because of that affiliation. Evaluate them the same way you would any other tool in this list. Tool availability, pricing, and features change — verify current details directly with each provider.

CRM and Business Operating System

Service Hubb AI — servicehubbai.com The CRM platform built specifically for service businesses, covering lead capture, pipeline management, AI voice and text agents, follow-up automation, review request automation, scheduling, dispatch, and 80/20 reporting. Built for the service trade workflow, not adapted from a generic sales tool. Pricing at time of writing: $297/month with a one-time $750 integration fee covering SMS compliance setup and full workflow buildout. *Referenced in: Chapters 11, 13, 16, 17, 18, 21, 31, 32*

Jobber — getjobber.com Full-featured field service management platform covering quoting, scheduling, invoicing, and client management. Strong mobile app. Widely used across service trades. *Referenced in: Chapter 21*

Housecall Pro — housecallpro.com Field service software with CRM, scheduling, payment processing, and customer communication tools. Popular in HVAC, plumbing, cleaning, and related trades. *Referenced in: Chapter 21*

ServiceTitan — servicetitan.com Enterprise-level field service management platform for larger operations. Deeper analytics and reporting than most alternatives. Higher price point.

AI and Quoting Tools

Haulers' Edge AI — haulinghubbb.com AI-assisted quoting tool for junk removal and hauling operations. Upload job photos, add notes, receive a price range and margin estimate. Built for the hauling trade specifically. *Referenced in: Chapters 1, 3, 33*

Claude — claude.ai Conversational AI for writing, research, analysis, and agentic workflows. Used for Claude Code research agents described in Chapter 33. *Referenced in: Chapters 30, 31, 33*

ChatGPT — chat.openai.com Conversational AI for drafting, research, and analysis. Strong general-purpose tool for the use cases described in Chapters 30 and 32. *Referenced in: Chapters 30, 32*

Google Ads and Paid Search

Google Ads — ads.google.com Paid search platform covering Local Services Ads (LSAs) and standard search campaigns. The primary paid acquisition channel for most local service businesses. *Referenced in: Chapter 14*

CallRail — callrail.com Call tracking platform that connects phone calls to the specific keywords, ads, and sources that generated them. Essential for measuring paid search ROI. Approximately $30–$100/month depending on plan. *Referenced in: Chapters 14, 21*

Google Local Services Ads — ads.google.com/local-services-ads Pay-per-lead format that appears above standard Google Ads. Requires background check, license verification, and GBP connection. Covered in Chapter 14. *Referenced in: Chapters 13, 14*

Website and SEO

Google Business Profile — business.google.com Free listing that controls how your business appears in Google Maps, local search, and AI search summaries. The single highest-return digital asset for most local service businesses. *Referenced in: Chapters 9, 13, 14, 34*

PageSpeed Insights — pagespeed.web.dev Free tool from Google that scores your website's load speed on mobile and desktop and identifies specific improvements. Target under three seconds on mobile. *Referenced in: Chapters 13, 14*

Google Search Console — search.google.com/search-console Free tool showing how your website performs in Google search — which queries drive traffic, click-through rates, and indexing status. *Referenced in: Chapter 13*

Google Analytics — analytics.google.com Free website analytics platform showing visitor behavior, traffic sources, conversion rates, and more. *Referenced in: Chapter 13*

Routing and Fleet

OptimoRoute — optimoroute.com Route optimization software for service businesses. Integrates job lists, time windows, truck capacity, and dump run requirements into optimized daily routes. *Referenced in: Chapters 19, 21, 31*

Route4Me — route4me.com Multi-stop route optimization platform. Handles larger operations and more complex routing scenarios. *Referenced in: Chapters 19, 21*

Google Maps — maps.google.com Free multi-stop route optimization for up to ten stops. Adequate for solo operators and small teams before dedicated routing software is justified. *Referenced in: Chapters 19, 21*

Samsara — samsara.com Fleet tracking and telematics platform providing real-time GPS visibility, driver behavior monitoring, and dispatch tools. *Referenced in: Chapter 21*

Verizon Connect — verizonconnect.com Fleet management and GPS tracking platform for multi-truck operations. *Referenced in: Chapter 21*

Payments

Square — squareup.com Mobile card readers, payment links, invoicing, and accounting integrations. One of the most widely used payment platforms for small service businesses. Processing fees typically 2.6–2.9% + fixed per transaction. *Referenced in: Chapter 21*

Stripe — stripe.com Payment processing with mobile readers, payment links, invoicing, and developer-friendly integrations. Similar pricing to Square. *Referenced in: Chapter 21*

Accounting and Expense Tracking

QuickBooks — quickbooks.intuit.com Small business accounting software covering P&L, balance sheet, cash flow statements, payroll integration, and bank account connectivity. The most widely used accounting platform for small service businesses. *Referenced in: Chapters 1, 21*

Wave — waveapps.com Free accounting software covering invoicing, expense tracking, and financial reporting. Strong option for early-stage businesses. *Referenced in: Chapters 1, 21*

MileIQ — mileiq.com Automatic mileage tracking app that runs in the background on your phone and logs business miles for tax deduction purposes. *Referenced in: Chapter 21*

Everlance — everlance.com Automatic mileage and expense tracking app with IRS-compliant reports. *Referenced in: Chapter 21*

Scheduling and Communication

Slack — slack.com Team communication platform with channels by topic, searchable history, and integrations with other business tools. Useful for multi-person teams. *Referenced in: Chapter 21*

Google Calendar — calendar.google.com Free shared calendar that works as a functional scheduling tool for solo operators and small teams before purpose-built scheduling software is needed. *Referenced in: Chapter 21*

Haulers' Edge Newsletter and Resources

Haulers' Edge Newsletter — haulinghubbb.com Weekly newsletter covering what's working in service businesses right now — marketing, operations, technology, and financial strategy. The ongoing companion to this book. Free to subscribe.

The tools, platforms, and pricing in this appendix reflect what's available at the time of writing. The landscape changes. For current recommendations, the newsletter covers updates as they happen.

www.ingramcontent.com/pod-product-compliance
Ingram Content Group UK Ltd.
Pitfield, Milton Keynes, MK11 3LW, UK
UKHW022029190726
13853UKWH00005B/2169